D1010610

Froth!

Mark Denny

THE JOHNS HOPKINS UNIVERSITY PRESS
BALTIMORE

Printed in the United States of America on acid-free paper
2 4 6 8 9 7 5 3 1

The Johns Hopkins University Press
2715 North Charles Street
Baltimore, Maryland 21218-4363
www.press.jhu.edu

Library of Congress Cataloging-in-Publication Data
Denny, Mark, 1953–
Froth! : the science of beer / Mark Denny.
p. cm.
Includes bibliographical references and index.
ISBN-13: 978-0-8018-9132-8 (hardcover : alk. paper)
ISBN-10: 0-8018-9132-9 (hardcover : alk. paper)
1. Beer. 2. Brewing. I. Title.
TP577.D476 2009
641.6′23—dc22 2008022646
A catalog record for this book is available from the British Library.

Special discounts are available for bulk purchases of this book. For more information, please contact Special Sales at 410-516-6936 or specialsales@press.jhu.edu.

The Johns Hopkins University Press uses environmentally friendly book materials, including recycled text paper that is composed of at least 30 percent post-consumer waste, whenever possible. All of our book papers are acid-free, and our jackets and covers are printed on paper with recycled content.

To friends from an earlier era—

John Hewitt, who taught me how to drink beer;
John Hardy, who taught me how to make it;
and my University pals, who taught me why

Contents

Acknowledgments

When I first pitched the suggestion of a beer-and-physics book to the Johns Hopkins University Press Editor-in-Chief, Trevor Lipscombe, he provided much encouragement. Later, during the fermentation stage, I was helped significantly by Horst Dornbusch, who influences beer and beerocrats on two continents, and who obtained for me special permission to raid the photo archives of the Bavarian Brewers Federation. The book has been brought into top condition at JHUP by copy editor Carolyn Moser and art director Martha Sewall. I thank you all.

In recent months my homebrewing has benefited from the generosity of John Rowling, who let me loose in his garden to pick hops. Cheers, John.

Froth!

Introduction

There can't be good living where there is not good drinking.
—Benjamin Franklin (1706–1790)

There can't be much amiss, 'tis clear,
to see the rate you drink your beer.
—A. E. Housman (1859–1936)

Beer is the most popular alcoholic beverage in the world. Estimates of worldwide annual consumption vary from 114 to 132 billion liters. Think of a lake 2 miles across and 30 feet deep.[1] Or, perhaps more apt, think of a giant beer glass half a mile high and a quarter mile across. I am talking about a lot of beer.

Before getting into the statistics a little more, I should tell you what this book is about. It may already have dawned on you that I am writing about beer, but there is more to my story than that. There are quite literally hundreds of books and Web sites about how to brew beer. Some of these are excellent (see the bibliography at the end of this book). Most fall into one of two categories, which I would characterize as "How-To Plus a Lot of Recipes" and "Beta-Amylase Influence on Maltose Production from 2-Row Grain." The first category is self-explanatory and runs the gamut from excellent to awful.[2] The second category consists of ultra-technical accounts of the brewing process

1. I wonder what waterfront (or beerfront) property values would be on Beer Lake. I can see it both ways, high and low.
2. I recall an early homebrew book written in England in the 1960s that included plaster of Paris as an ingredient for one beer recipe.

and seems to be written for professional brewers, academic researchers, or the geek end of the homebrew market. Both can provide interesting and useful information for the homebrewer, and some books (such as Wheeler's *Home Brewing*) successfully combine elements of both categories. My book is unique, to the best of my knowledge, in that it unites brewing with accessible physics. You are not holding in your hands a recipe book or a Ph.D. thesis, but if you are interested in beer, and about how science and technology impact the production of your favorite tipple, then you will find much to engross you in the following pages.

Math analysis and beer tend not to go together in the literature. I am a physicist by training and a homebrewer by inclination. Inevitably I have, over the years, applied my knowledge of physics to the science of brewing. The results are, I believe, better brews and a better understanding of the brewing process. So, herein you will find out about beer and brewing in general, and about how to homebrew good beer, in particular. My science slant will be evident: math will occasionally be introduced, but the text is written so that, if you wish, you can glide over the equations without missing out on anything.

Mother Nature speaks mathematics, but most people don't, so I am well aware that the text should be stand-alone. Those of you who happen to be interested in the math as well as the beer (and in my experience most mathematicians, physicists, engineers, and science students are partial to both) will find that the technical aspects of brewing lend themselves well to mathematical analysis. More about all that later: here, I would like to return to the statistics of beer, after a necessary paragraph about units.

In table I.1 you will find conversions between a few of the many and varied units that have evolved over time, and in different countries, for the measurable quantities used in making beer. The diversity of units can be confusing without such a handy table to effect a translation. Throughout the book, when I mention "gallon" I mean a U.S. gallon, which is not the same as an imperial (English) gallon. On the other hand, the physicist in me likes decimal units, and so I will perform calculations using liters, kilograms, degrees Celsius, etc. When I feel like it, I may convert to more familiar units in the text; otherwise, please remember the page number of table I.1 and refer back to it, if puzzled. All of the units in this table are used in the text. I have tried to

Table I.1 Conversion factors pertaining to beer

Temperature	100°C	212°F	
	70	158	
	60	140	
	35	95	
	25	77	
	15	59	
	0	32	
Density[a]	1.06 SG	1060 OG	6.3% abv
	1.05	1050	5.3
	1.04	1040	4.2
	1.03	1030	3.2
Weight	1 ounce	28 grams (gram weight)	
	2.2 pounds	1 kilogram	
Proof	1° U.S.	0.87° English	0.5% abv
Volume	1 U.S. gallon	0.83 imperial gallon	3.7 liter
Power	1 kW	239 cal s^{-1}	

[a]SG = specific gravity (density / density of water), OG = original gravity, a unit favored by brewers, abv = alcohol by volume.

reduce the information in table I.1—and the different units used in the text—to a minimum, leaving out, for example, some of the European units for alcohol content and some of the strange historical weights and measures, in particular the multitude of names for different bottles and barrels. Here is just one example to provide a flavor of the variety of beer and ale containers (wine is different): there are 54 imperial gallons in a hogshead, 36 in a barrel, 18 in a kilderkin, 9 in a firkin, and 4½ in a pin. Of these units, only the barrel (abbreviated *bbl*) is widely used today in the brewing industry, though some retailers sell beer by the pin.

Now for some more of those telling statistics about beer. Per capita, the Czech people drain Beer Lake faster than any other nationality, as you can see from figure I.1. The nations that swill the most beer are listed in figure I.2. There are perhaps a couple of surprises that emerge from these two graphs.

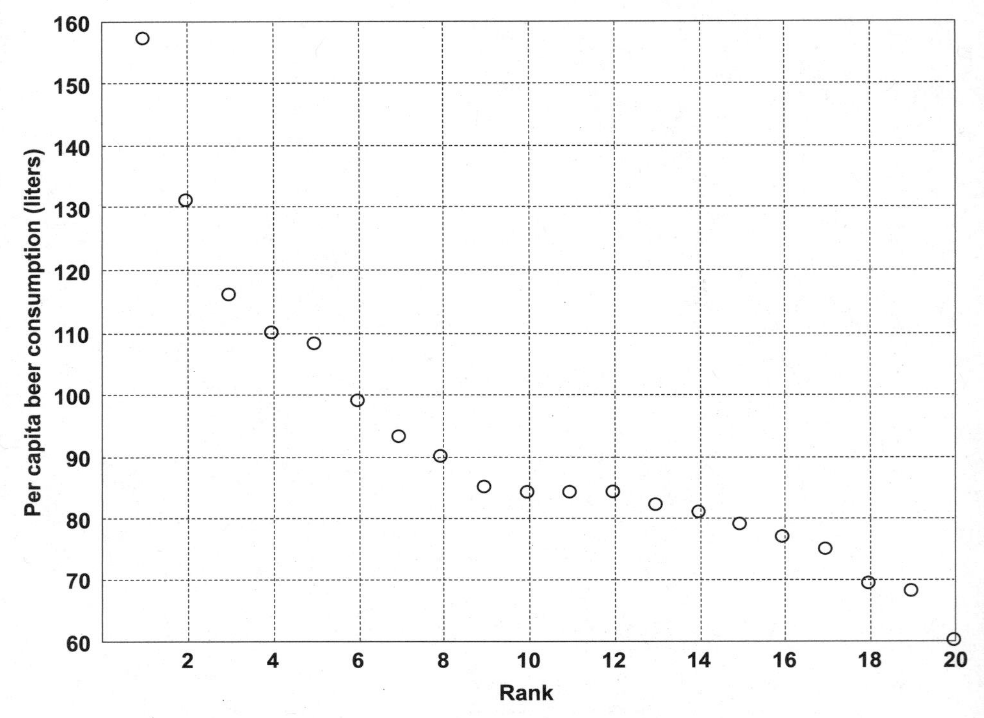

Figure I.1. Annual per capita beer consumption by nation: (1) Czech Republic, (2) Ireland, (3) Germany, (4) Australia, (5) Austria, (6) United Kingdom, (7) Belgium, (8) Denmark, (9) Finland, (10) Luxembourg, (11) Slovakia, (12) Spain, (13) USA, (14) Croatia, (15) the Netherlands, (16) New Zealand, (17) Hungary, (18) Poland, (19) Canada, (20) Portugal. *Source:* The Kirin Brewing Co.

The presence of Spain and Portugal in the top 20 for per capita beer consumption may raise an eyebrow or two—except in Spain and Portugal. We might expect wine to dominate in these southern European countries, but, it seems, the Iberians like their beer as well. In fact, given the hot summers in that part of the world, and the chilled lagers brewed in Spain and Portugal, we can readily understand the appeal of a cool brew (ditto Australia, Mexico, and the United States).

The city which claims the greatest per capita consumption of beer is Darwin, in northern Australia. Here, a sweltering climate, a long tradition of beer drinking, and a macho culture combine to produce a beer consumption rate of 504 pints (233 liters) per person per year. That is about equal to 10 U.S. pints per week, for every man, woman, and child. Gulp.

Another surprise: the United States has been overtaken since 2004 as the top beer-guzzling nation. The taste for beer has reached China, and the world's most populous country has now taken over pole position, as you can see in figure I.2. Worldwide, beer consumption has increased annually for each of the last 19 years due, at least in part, to increased summer temperatures. In the United States 87% of alcohol consumed is via beer. Forty-three percent of all the beer drunk is swilled by 10% of the beer drinkers. Consumption is not strongly correlated with income, but the type of beer consumed does vary with economic status. High-income earners are more likely to drink light beers or imported beers.

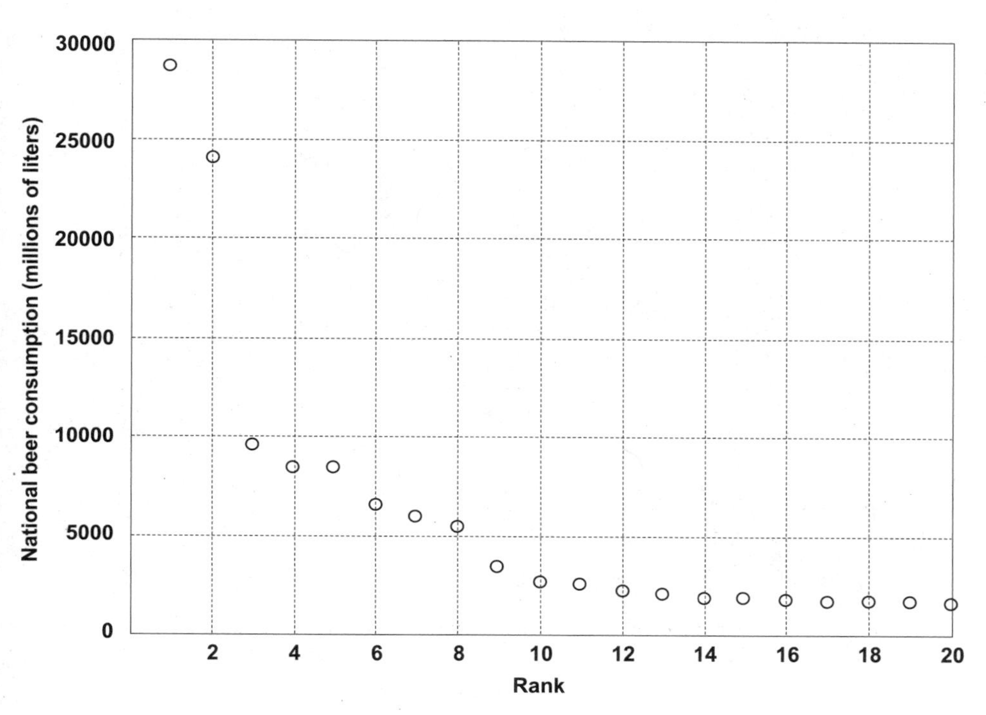

Figure I.2. Total annual beer consumption by nation: (1) China, (2) United States, (3) Germany, (4) Brazil, (5) Russia, (6) Japan, (7) United Kingdom, (8) Mexico, (9) Spain, (10) Poland, (11) South Africa, (12) Canada, (13) France, (14) South Korea, (15) Czech Republic, (16) Ukraine, (17) Italy, (18) Australia, (19) Colombia, (20) Thailand. *Source:* The Kirin Brewing Co.

Figure I.3. You may have drunk only one bottle of Belgian Duvel beer, but this is what you see.

Table I.2 The main beer-producing nations (output in millions of barrels)

Country	1991	1996	2000	2001
USA	145.0	144.2	142.7	143.5
China	51.2	99.7	134.7	138.7
Germany	72.1	69.8	67.5	66.3
Brazil	39.7	54.1	50.5	51.0
Japan	41.9	42.0	43.8	43.9
Russia	NA	12.3	33.5	38.5
Mexico	25.2	28.8	35.3	36.0
Britain	36.4	35.5	33.8	34.7
Spain	16.2	15.1	16.1	16.9
Netherlands	12.0	14.4	15.3	15.4

Source: British Beer and Pub Association.

In the chapters to follow, I discuss the different types of beer originating in different parts of the world. These various brewing traditions give rise to different strengths (alcohol by volume, or *abv*) of beer around the world. Thus, in England the beer and ale has an average strength of 4.4% abv, whereas in the United States (and most of the rest of the world), the favored lager style of beer is usually 5.0% abv. Belgium, with its own unique, bizarre, and delicious tradition of beers (e.g., fig. I.3), tops out with an average alcohol content of 8.0% abv.

So much for consumption: what about beer production? In table I.2 you can see the world's largest contributors to Beer Lake. The United States still heads the table, but China is poised to take over the lead. The output of the traditional big producers of Europe (Germany, Britain) is static or declining as other types of beverage become more popular. In Russia a change in government policy (perhaps in a drive to reduce vodka consumption and alcohol abuse) and investment by foreign brewing companies has resulted in a big increase in beer production over the last decade.

Within the United States, 82% of beer is produced by three dominant breweries: Anheuser-Busch (52%), Miller (19%), and Coors (11%). The many and varied microbreweries are relatively miniscule, but increasing in number, for reasons that I will describe.

I hope that these bald statistics have gotten across to you the undeniable fact that the world likes beer, and likes it a lot. I am reminded of a drinking friend of mine from university days. Mikel said one evening, in a pub in Edinburgh, Scotland: "I like drinking a lot!" Succinct and to the point, you might suppose. Mikel, however, retained the scientist's precision of thought even at that late stage of the evening and realized that his statement was ambiguous. Did he mean that he was very fond of drinking beer (we drank little else in those days), or did he mean that he liked drinking large quantities of beer? To provide clarification, Mikel stood up, perhaps unsteadily, and said: "I like drinking a lot, a lot!" Now I do not wish to be accused of encouraging alcohol abuse, but a little tongue-in-cheek humor is appropriate in this context, I feel.[3]

Despite the millions of Mikels in the world, however, all has not been well in the international beer community over the past 30 years or so. Many beer connoisseurs came to feel that the unstoppable increase in the size of big breweries was leading to a decline in beer quality. The reasons for this perceived fall-off in big-brewery beer, and the consequent flowering of microbreweries and of homebrewed beer, will be aired in the first chapter.

In chapter 2, I will tell you how I make my homebrew. To whet your appetite, please consult figure I.4. I have been brewing for about 15 years and have, by adopting a scientific approach (i.e., enlightened trial and error, plus some math analysis), pared the process down to the simplest possible method—though I continue to tweak the method with experimental refinements from time to time. I adopt the full-mash infusion approach and prefer the English style of beer (top-fermenting—e.g.,

3. Thus: Government Health Warning labels should be read in moderation. Avoid binge reading (more than five labels at one time). Read responsibly.

I cannot resist telling the following story, every word of which is true, I swear. My wife and I emigrated from Scotland to Canada a few years ago, and as part of the Canadian immigration requirements we were obliged, while still living in Scotland, to visit a designated medical practitioner for a health check. The doctor who examined me was typical of the Scottish medical profession. He was a Scot (with and without the *c*) who possessed a bright red nose and a feisty disposition. He was filling out a form for the Canadian authorities and asked me how much I drank. I replied truthfully, "About twenty pints of beer per week." I felt a little guilty, since this amount was perhaps a little more than is medically beneficial, but I need not have worried. The good doctor screwed up his face, as if he had just swallowed a wasp, and said, "Twenty pints? That's not nearly enough. I'll write doon 'light drinker.'"

Figure I.4. There is nothing quite as satisfying on a hot summer day as one of these. One of the pleasures of homebrewing is that we can adjust recipes and techniques to customize a beer—to tailor it to circumstances and personal taste. So, for summer I brew a light-bodied, hoppy brew that retains flavor when refrigerated. My winter beers are darker and more full-bodied and are served at cellar temperature.

IPA, mild, stout) rather than the continental European style of lager (bottom-fermenting—e.g., pilsner). My description will be placed in context, in that I will describe the main differences between small-scale homebrewing and large-scale commercial brewing, and between beer and lager brewing.

The third chapter describes a more theoretical approach to the study of beer. I will show how the population of yeast that is pitched into a batch of homebrew grows exponentially at first, and then suffers a catastrophic population crash as food resources are used up. There are practical consequences for the dedicated homebrewer that follow from this calculation, as we will see.

During the entire process of making beer, from mashing to bottling, the brewer is anxious to maintain the correct temperature. In chapter 4, I will discuss the importance of temperature control and share my calculations describing various aspects of beer thermodynamics, mostly from the perspective of a homebrewer.

Bubbles are very important to the esthetics of beer. They have been the subject of numerous scientific papers; research on this topic has even received an Ig Nobel Prize. Bubbles arise during fermentation, as well as during the pouring of a glass of beer, and they continue to rise (and fall) after the beer is poured and the froth has formed a head. All this and more in chapter 5.

Chapter 6 looks at beer as a fluid, rather than as the Amber Nectar, the Elixir of Life, the golden (or brown, or black, or even pink or orange) liquid worshipped by beerophiles.[4] The distribution and dispensing of beer presents problems that have shaped the way beer is made and have also shaped the way that beer is presented to the connoisseur at public institutions dedicated to the appreciation thereof (pubs, to you). The final chapter takes a sideways glance at brewing as an application of science to everyday life. Many years ago an inebriated botanist once opined to me that "it's beer puts the fizz into physicists," and he may have been right.

So, this book is about the evolution of beer and the manufacture of beer on both large scale and (particularly) small scale. My story has a technical account—including math, which can be read around if algebra makes you squirm—of the many ways in which physics enters Zymology.[5]

Finally, because they cast interesting or amusing sidelights upon the subject of this book, if you haven't already discovered them I direct your attention to the . . .[6]

4. I don't know if *beerophile* is a real word or not, but it should be, and I will use it freely in this book.

5. The study of fermentation.

6. . . . footnotes.

One

The Evolution of Beer

He was a wise man who invented beer.
—Plato (c. 429–347 BC)

A fine beer may be judged with only one sip,
but it's better to be thoroughly sure.
—Czech proverb

This historical account will begin quite generally as a mild, well-balanced, and tasty (I hope) instruction upon the origin and development of beer in western Europe up to the turn of the twentieth century. Following an intermission, during which you will discover where I am coming from, and where I am going to, you will find a rather frothy history (a touch acidic) of developments from 1900 to the present day, mostly in the United States. My aim here is to provide a backdrop that places our modern brewing (and appreciation) of beer in context and to show you why the beer world is moving in the direction we see. Ultimately you will find out why many beerophiles brew their own beer nowadays, and so this introductory chapter will naturally lead into the subject of how to brew beer at home, which interesting topic forms the substance of chapter 2.

For you to appreciate the history of beer more fully, I need to recap briefly the basics of traditional beer making. Details are provided later; here we require only a broad-brush summary. Beer is a slightly alcoholic beverage—typically 4%–5% alcohol by volume (*abv*)—that is made by fermenting cereals. Thus, grain is malted (gently heated until it partially germinates), before being cracked open and mixed with water (*mashed*). The resulting starchy liquid is then boiled and allowed to cool. Yeast is added. The yeast multiplies, feeding off certain sugars in the starch, and eventually turns these into alcohol; this is the *fermentation* process. During fermentation, certain adjuncts may be added to influence the flavor and character of the beer. Then the yeast dies, or goes into a kind of stasis, and settles on the bottom of the fermentation vessel. The green (i.e., young) beer is drawn off the sediment and stored, in airtight containers, in one of several different ways. During this storage period there occurs a further, slow fermentation that serves to carbonate the beer—make it fizzy. After a certain maturation period, the beer is then drunk, to the satisfaction of everyone concerned.

The brewers of old did not know about yeast, as we will see, and so to them the transition from starchy water to foaming beer must have seemed magical. Nowadays brewers make use of two general types of yeast. Lager yeast ferments while sitting at the bottom of the fermentation vessel and works best at low temperatures. Ale or beer yeast sits on top of the starchy liquid (the *wort*, pronounced "wurt") and prefers higher temperatures.

I will now send you, armed only with these basic facts and a curiosity about our favorite tipple, back in time about 10,000 years, to the Middle East.

THE DAWN OF CIVILIZATION

Charlie Bamforth, professor of brewing at the University of California, has made the extravagant claim that "beer is the basis of modern static civilization" (Mirsky). German author and beer guru Horst Dornbusch has written an article with the title "Beer: The Midwife of Civilization" (Dornbusch). It is amusing to think that some prehistoric man was laboring away hunting mammoths one day and thought to himself, "Yikes, I need a beer—let's invent civilization," but I suspect that this is not what Bamforth and Dornbusch had in mind. The idea is that beer requires a cereal crop, which requires agriculture, which requires an organized, sedentary culture. Previously, people had grouped together in nomadic hunter-gatherer bands. So the suggestion is that the thirst for beer drove early man to settle down and develop farms, towns, roads, and the infrastructure of organized agriculture.

Maybe this view is extreme, though I find it rather appealing, but it is certain that beer making is among the most ancient manufacturing arts known to man. The brewing of beer, in one form or another, is as old as the baking of bread. Indeed, one may be a by-product of the other. Some historical sources guess that beer arose when bread was dunked in water, or became wet, and was then left for some time.[1] Yeast in the bread led to fermentation, and the resulting wet, moldy bread will have been alcoholic. Other historians think that prehistoric nomads made beer from wild grain and water before learning how to bake bread. Or maybe an accidental soaking of grain, warmed in the sun and fermented by wild yeast, led to intoxicating liquor which man, being man, decided he wanted more of. (James and Thorpe, and the Beer History and German Beer Institute Web sites, listed in the bibliography, provide details about the origin of beer.)

However it happened, it happened about 10,000 years ago in the Middle East, and spread out from there to other parts of the world. Different cereals were used in different parts of the globe. In more recent times, hundreds of beers around the world have been brewed from locally available products: barley, wheat, and rye in Europe and the Middle East; cassava, millet, and sorghum in Africa; rice in eastern Asia; cactus in Mexico; corn and sweet potato in the Americas. However, barley has become the grain of choice for most beer makers today, and this preference is an ancient one. The Sumerians, in modern Iraq, were the first people to brew beer (or, rather, the earliest beer records unearthed by historians are Sumerian), and they used barley. By 6000 BC the Babylonians were also at it, and then the ancient Egyptians. Reliefs on stone tombs show partially germinated barley being crushed, mixed with water, and then fermented: clearly the process being illustrated is the brewing of beer. These ancient civilizations obviously valued beer and developed it to the extent that many distinct brews were made. The Sumerians produced about 10 varieties. Babylonians are known to have made at least 34 varieties of beer; we know this because, around 4300 BC, they produced clay tablets detailing beer recipes. Beer production

1. Today the eastern European beer *kvass* (drunk in Russia and the Baltic states) is made from fermented rye bread.

was a state monopoly in ancient Egypt, with strict rules on the methods of production, because beer offerings were part of the pharaohs' religious practices.

Brewing was important enough for Egyptian brewers to have their own special hieroglyph. It is said that if an Egyptian man offered a lady a sip of his beer, and she accepted, then they were betrothed. The Sumerians had a goddess of brewing. Their famous Gilgamesh epic, a heroic poem that is perhaps the oldest written story on earth (set down before 2500 BC), refers to beer as a product of civilization, separating "cultured man" from barbarians. The great Babylonian priest-king Hammurabi, who united all of Mesopotamia around 1750 BC, left a code of laws that is reckoned to be the oldest in the world. It includes a daily beer ration that depended upon social status, with more important people getting more beer. Also, Hammurabi's Code includes a law governing the pricing of beer, with an appropriate punishment (drowning) for tavern keepers who sold short measure.

So what would the beer of these three ancient civilizations—Sumeria, Babylonia, and Egypt—have been like? Some reckon that it would have been a thick sludge, like porridge. It may not have been quite as thick as porridge, since it was drunk through a kind of straw. The purpose of the straw may have been to filter out bitter-tasting "floaters." The liquor was probably quite sour. Ancient beers are described as being dark, pale, red, with and without a head, and so on. There must have been different additives to produce these different effects.

Beer also formed a part of ancient Hebrew culture.[2] It arose (presumably independently) in ancient China and among the Incas of pre-Columbian South America. Whether drunk as part of religious ritual, or used in lieu of pay for workers, or consumed for pleasure, beer and beer production were well-established aspects of civilized culture by the time that they reached Europe.

2. "Shebrew" might be more appropriate in this context, since women were responsible for brewing beer in much of the ancient world. The Babylonian brewers were priestesses. The dominance of women brewers would change in Western Europe during the Dark Ages, when brewing was taken over by monastic orders, but women would regain ascendancy later, as we will see.

BEER WASHES UP ON EUROPEAN SHORES

It is thought that beer spread from Egypt to classical Greece and from Greece to Rome. We tend to think of the denizens of these two European civilizations of classical antiquity as wine drinkers, but in the early days they imbibed beer. Indeed, when Caesar crossed the Rubicon he toasted his officers with beer. (Our word "beer" comes from the Latin *bibere*—to drink.[3]) The expanding Roman Empire introduced many benefits of civilized living to the northern barbarians, but beer was not one of them, for the simple reason that the northern barbarians already had a long tradition of brewing beer (so they can't have been all that barbarous). The knowledge of brewing had spread from the Middle East to northern Europe along the river Danube, or across the Mediterranean to southern France, and then northwards. With the development of viniculture around the Mediterranean, beer became less popular and eventually came to be seen as a barbarian drink. The Roman historian Pliny the Elder reported that beer had been the drink of choice before winemaking spread across the Roman world, and both he and Tacitus (both first century AD) tell of ale-drinking Celtic, Nordic, and Germanic tribes.[4] Tacitus goes on to describe beer in these negative terms—a sign of the changing times: "To drink, the Teutons have a horrible brew fermented from barley or wheat, a brew that has only a very far removed similarity to wine."

The migration of beer from southerly regions (the Middle East and the Mediterranean) to northern Europe ensured its survival. Wine became ever more popular and vineyards took over from barley fields in the centers of civilization, pushing barley to the northern fringes. Both wheat and barley were grown in the south, but barley was found to grow better in the cooler northerly climes. Wheat made better bread, but barley was the more suitable cereal crop for beer. This is because barley could be stored more readily after harvesting, for long periods—

3. The Spanish word for "beer," *cerveza,* comes from the Latin *cerevisia,* which in turn comes from Ceres, the ancient Greek goddess of agriculture.

4. Pliny, *Natural History:* "The perverted ingenuity of man has given to water the power of intoxication. Where wine is not procurable, western nations intoxicate themselves with a kind of moistened grain."

a key advantage in the days when brewing could be done only in winter, for reasons I will soon make clear.

In Roman Gaul (modern France) the women brewed beer as a cottage industry.[5] Further east, the German tribes already had a long familiarity with beer; archaeologists have discovered beer amphorae (large earthenware vessels for storing liquids) dating from 800 BC. These were found near the German city of Kulmbach, which is still a brewing center today. At the height of the Roman Empire, when Romans and Germans were eying each other nervously across the Rhine, there was significant commercial trade in beer. Germans considered intoxication to be divine (perhaps many still do), and their beer was brewed for religious observance as well as for enjoyment. Another feature they shared in common with earlier brewing traditions is that their brewers were female. The Romans never conquered Germany, but they did get most of Britain. Ale was well established in England at the time that the Romans first crossed the Channel, in 55 BC.[6]

In those days, all beer was brewed from a varying mixture of wild top-fermenting ale yeast and bottom-fermenting lager yeast. (These two varieties would not be isolated until the nineteenth century.) The barley was malted and kilned (heated) over open wood fires, which led to a smoky flavor and dark beer, a feature of beer that would last until the 1840s. Brewing at this stage was a crude, poorly understood, and very inexact science—indeed, not a science at all, but an art. Another characteristic of ancient brews that would persist up to the nineteenth century is this: beer was safer to drink than water. This brutal fact may by itself account for the popularity of beer in historical times. Waterborne diseases such as cholera would periodically break out in epidemics that would kill people by the thousands. You could not tell whether the water you were drinking was contaminated with deadly cholera or was perfectly fresh—the taste was the same. Beer that was contaminated with bacteria, however, tasted vinegary. Plus, the brewing process itself killed off any undesirable bacteria that may have been lurking in the

5. In Gaul malted barley was called *brace,* from which the modern French word *brasseur,* or brewer, is derived.

6. Indeed, one of the reasons that the Romans wanted Britain is because the ancient Britons produced a large amount of grain.

water used for brewing (we will see why in the next chapter). So, beer that tasted good had no harmful bugs in it. Harmful bugs soured beer, and this was easy to detect. Drinking beer was good for you.

FROM DARK ALE TO BEER, FROM DARK AGES TO ENLIGHTENMENT

The Romans left northern Europe like a receding tide, but beer remained. Barbarism returned; these were the Dark Ages during which, in many ways, civilization was put on hold for a millennium. The one (possibly the only) unifying and improving force left in western Europe was that of the monasteries. Christianity had been declared the state religion of the dying Roman Empire, and it survived barbarian invasions that destroyed Rome, defining "Christendom"—for better or worse—until the Enlightenment brought humanity kicking and screaming into the modern world.

But let's cut to the chase: what did the monks do for beer? First, I need to clear away a source of confusion: *ale* or *beer*? In fact the distinction is blurred, and the linguistic origins are as murky as the ancient beer. Our word "ale" comes from the Old German word *öl*, which describes the beverage drunk by Germanic tribes in the Dark Ages. Nowadays many people refer to beer and ale interchangeably, while others draw a distinction based upon the use of hops. I will have a lot to say about hops later on, but they have not yet entered our history. Let us adopt this increasingly common convention: *ale* refers to beer that contains little or no hops, whereas *beer* refers to the hopped product. I emphasize that this distinction is a modern convention and may not be historically justified. Nevertheless it is useful, in that it emphasizes the important role that hops will play in the history of beer evolution, and it permits an easy distinction between ales brewed in ancient times and beer brewed in recent centuries. So far in this book I have used the generic term "beer" to describe all fermented cereal liquors, but henceforth I will be more specific. "Ale" will refer only to the unhopped brew made in ancient times before yeast was known about (and so ales may have been fermented by either top- or bottom-fermenting yeasts, or both, depending upon the time of year—recall that they thrive at different temperature ranges). I will continue to use the word "beer" in the more general sense but also will use it to refer specifically to top-fermented brews that are hopped; you should be able to tell from context which meaning is intended. Later we will come to "lagers," which are, strictly, bottom-fermented and hopped brews.

At the beginning of the Dark Ages (which description may be applied to the ale as well as to the times) ale was made by throwing half-baked bread into water and letting nature do the rest. Ale was made at home by an "alewife"[7] who made use of whatever cereals and adjuncts were available to concoct her brew. Starch could be extracted from wheat, rye, and oats as well as barley, and even from peas or beans. Additional flavors were derived from wild herbs such as bog myrtle, juniper berries, rosemary, yarrow, and hop flowers, which grew wild over much of central Europe. Other additives included blackthorn, oak bark, wormwood, St. John's wort, and henbane (a hallucinogen).

As the monasteries gained power (they were the only unifying force, since the feudal system naturally led to divisions among secular authorities), they began to take over brewing. They owned a lot of land, upon which cereals were grown and wild herbs were harvested. They were self-reliant. The monks were required to fast, but these fasts did not include liquid nourishment such as ale.[8] So, you can see how monasteries came to take over brewing.

The monks improved brewing practices, refining the brewing process over centuries. A few monastic breweries still exist today.[9] Brewing has several patron saints: St. Augustine, St. Luke, St. Nicholas (a.k.a. Santa Claus). In many places, the dominance of monasteries was solidified by legislation: nobody else was allowed to brew ale. This measure suited the monasteries just fine because they had learned to make money by providing outlets for their beer to the general public. For

7. Our word "bridal" comes from *bride-ale*, a product brewed to celebrate weddings. Brewing terms used in English, such as "mash" and "wort," originate from the Anglo-Saxon that was spoken in England during the Dark Ages.

8. The attitude of medieval monks to their nutritious "liquid bread" was expressed as "*liquida non frangunt ieunium*" (liquids do not break the fast). Drinking ale on an empty stomach must have produced many a drunken monk. I have to say it: bringing ale into religious observance seems to me to be a classic case of mixing spirits.

9. In all the European countries with a strong brewing tradition: Belgium, Czech Republic, Germany, England, and others.

example, in German lands each monastery had its own *Kloisterschenken,* or taproom, where passers-by could take away with them some well-brewed ale (for a price).

Eventually, secular authority got its act together, and the power of the church waned, as feudalism itself waned. Lords and city burghers both eyed the lands and revenues generated by monasteries, and a slow shift of power from church to state came to influence the history of beer (actually, still ale at this stage—we are now in the Middle Ages, say AD 1300). For example, in Bohemia "good" king Wenceslas persuaded the pope to revoke laws that banned brewing outside monasteries. Brewing guilds were established. Commercial brewing increased, and ale was traded widely (see fig. 1.1). Bavarians imported Czech beer from Budejovice, which they called *Buddweis* (hence Budweiser—see fig. 1.2) and from Plzen in western Bohemia (hence, *Pilsner* and *Pils*). German beer was exported to other countries via the Hanseatic League, a consortium of trading ports in medieval Europe. Thus, the city of Bremen exported ale to Scandinavia, Holland, and England. Hamburg became a major brewing center (with over 600 breweries in AD 1500) and exported ale to other Hanseatic ports.

At this time, and as a consequence of increased trade, ale began to evolve into beer. Hops had been utilized to flavor beer for several centuries—for example, they had been cultivated on Czech lands since the ninth century—but they were now increasingly added to ales because of their preservative value.[10] The female hop flower contains certain oily acids, as we will see, and these acids deter bacterial growth as well as add flavor. So, hopped ale (that is, beer) kept better than unhopped ale and thus could be transported over longer distances. For this reason it made good commercial sense for the brewers of continental Europe to hop their brews. In England it took a few more centuries for hops to catch on (hence the word "ale" still is associated with English beer); when hops were eventually added to English brews, it would be for historical reasons that had little to do with preservation.

10. As well as preserving beer, hops preserved women. Up until late in the sixteenth century, "beer witches" had been burned for spoiling brews. That is to say, women were done to death by tipplers whose beer was not up to scratch. The rise in hop use, and the consequent drop in spoiled beer, killed off this bizarre and barbaric practice.

Figure 1.1. A brewery in the Middle Ages.

Figure 1.2. Budweiser Budvar. A rich taste of malt and plenty of Saaz bittering hops distinguish this substantial and traditional Czech lager from its modern American namesake. There is a protracted international legal dispute over the ownership of the name *Budweiser.*

Another consequence of increased trade, arguably, was the rise in Germany of beer ordinances for quality control. There were dozens of such ordinances prior to the famous Bavarian *Reinheitsgebot,* or "purity law," of 1516, the oldest food regulation still applicable today.[11] This law stated that beer could be brewed only from malted barley, water, and hops. No additives (peas, beans, soot, wild herbs, etc.) were allowed. Reinheitsgebot law led to beer of higher and more easily controlled quality, brewed consistently.[12] Consistency of character and taste

11. Apparently the German purity law contravenes current European Union legislation, so that now Germans can import foreign beers that do not comply with this law.
12. Another reason for the law was to prevent competition with bakers for purchasing wheat and rye. One bad consequence of the Reinheitsgebot law is that many of

is important for a commercial product. At about this time, brewers were learning to differentiate between winter brewing and summer brewing. They did not yet know that the differing characteristics of winter and summer beers were due to different yeasts (cold-loving lager yeast and warmer ale yeast). Half the flavor of ales, beers, and lagers comes from the yeast, so it is not surprising that the two main groups of brewer's yeast produce different characteristics among their brews. A fuller understanding would come later, in the nineteenth century; meanwhile, Czech and German brewers were learning to *lager* (store, usually underground, where the temperature is lower) their brews for several months (favoring bottom-fermenting yeast). They had already learned that new brews fermented more quickly when some foam from a previous brew was added to it. We now know that this foam contained top-fermenting yeast. So, empirical brewing practices were giving rise to brews that were differentiated according to yeast type—lager and beer.

Hops were widely adopted in England reluctantly, in the seventeenth and eighteenth centuries, though they had been imported from Holland since the 1400s. Particularly during the eighteenth century, England and France were at each other's throats constantly. These were no bar-room brawls, but major wars involving many nations fought for the control of two continents.[13] The main results were that (a) the British Empire was established and (b) English ale was hopped. The historical connection between wars and hops goes like this:

1. England's wars with France cost a lot of money, and so
2. English taxes were raised, including ale tax, and so
3. ale became too expensive for many tipplers (bootleg gin became the scourge of London—see fig. 1.8 below), resulting in reduced sales, and so

---

the top-fermenting beers in northern Germany were suppressed; these beers usually required adjuncts. Today less than 15% of German beer is produced using top-fermenting yeast. Wheat beer also was *verboten* under the Reinheitsgebot law.

13. The eighteenth century alone saw the War of the Spanish Succession, the War of the Austrian Succession, the Seven Years' War (the North American part of which is the French and Indian War), the American Revolution, and the French Revolution.

4. brewers reduced the alcohol content of beer (because tax depended on alcohol content), and so
5. ale did not keep as well (since alcohol is a preservative), and so
6. hops were added, as a preservative, creating English beer.

During this period, the beer traditions of northern European countries became established. I can best describe these to you, and explain the next (crucial) phase of beer development, in a separate section.

BEER, PALE ALE, LAGER, AND INDUSTRIALIZATION

There are three main strands of European beer development, which here are labeled Czech and German, English, and Belgian. The evolution of European beer in the eighteenth and nineteenth centuries is conveniently encapsulated under these headings.

Czech and German Beer

I have lumped together the beer traditions of the Czech and German peoples (though both might object) because they are similar and because they have had significant mutual influence.

Throughout the eighteenth and nineteenth centuries, brewers' understanding of the brewing process increased, at first empirically (through the intelligent application of trial-and-error experimentation) and later through analytical understanding of the biochemical processes that underlie fermentation. The different effects of top-fermenting and bottom-fermenting yeasts were appreciated, and brewers learned how each brewing process was best practiced. Thus, without understanding yeast they learned that bottom fermentation worked best between 39°F and 48°F (4°–9°C), whereas top fermentation worked best between 62°F and 69°F (17°–21°C), and so, in the age before refrigeration, the former worked best in winter and the latter in summer. Bavarians (in the south of Germany and influenced by the Czechs) decided that they preferred the winter style of beer and banned summer brewing. This step led to a north-south differentiation among German beers that persists to this day.

Technological developments outlined below led to the kilning of malt by indirect hot air flow, rather than by direct heating via burning wood. This industrial revolution innovation changed the character of beer because wood smoke was removed from the wort; beer became paler and less smoky. At the same time in Germany the old brewers' guilds were being eroded by competition. Consequently, a trend of amalgamation and expansion developed that has continued and spread worldwide to the present day. For example, in Munich there were 60 breweries in 1790, but by 1819 these had combined into only 35 (larger) breweries. By 1865 the number had reduced to 15. Commercial competition led directly to innovation: in 1843 Balling, a Bohemian brewer, introduced the hydrometer, an essential tool of modern homebrewers that I discuss in chapter 2. In 1860 a German, Carl von Linde, perfected commercial refrigeration (first tested in a Munich brewery); the consequences for beer production and distribution were massive, as we will soon see.

German and Czech beers were, and are, very local, with each region developing its own characteristic style. When Germans go to a pub and order beer, they usually ask for a style, rather than a brand. Here is a partial description of the major Czech-German styles of beer.

Pils or *Pilsner*. This style has been much imitated (often indifferently), and today accounts for 90% of all beer sales worldwide (70% of sales in Germany). Originally *Pilsner Urquell* from Bohemia (made with the soft water of Plzen), this is a clear, hoppy, dry, and quite bitter lager.

Altbier. This copper-colored lager with a dry finish and medium body is characteristic of Düsseldorf. (See fig. 1.3.)

Bockbier. A strong winter lager from Munich that has a considerable following today. A heavy, malty, rich, barley-wine type of lager. (See fig. 1.4.)

Dunkel. A dark lager from Bavaria. Malty and lightly hopped.

Helles. A light beer (though most definitely *not* in the current American sense of the term), a straw-blond Munich lager. Dry and subtle. (See fig. 1.5.)

Kölsch. A blond native of Cologne. The German version of English pale ale.

Weissbier. A style of yeasty Bavarian wheat ale. Mildly hopped and complex. (See fig. 1.6.)

Figure 1.3. *Altbier* means "old beer," and this venerable brew has long been associated with the German city of Düsseldorf. The *grist* of this beer (the combination of grains that constitute the mash) is formed from Pilsner malt plus a small amount of black (well-roasted) malt, giving a copper color and a dry, malty flavor with a hint of caramel. A mixture of traditions in some ways, this splendid beer is top-fermented but also lagered.

Figure 1.4. Bockbier, a strong, dark winter lager. Photo courtesy of the Bavarian Brewers Federation, Munich, Germany.

Figure 1.5. Helles lager. This south German brew is less hoppy and more malty than Pils. The best-known Helles brand outside Germany, *Loewenbraeu* (also spelled *Lowenbrau* or *Löwenbräu*), is perhaps not the best of its type.

Figure 1.6. Weizen, or Weissbier. Photo courtesy of the Bavarian Brewers Federation, Munich, Germany.

English Beer

Hops, once they were accepted by English brewing practice, combined with the industrial revolution, changed the earlier ales that had been produced for centuries and resulted in modern "pale ale" and English "bitter" beer. But I am getting ahead of the story, which begins with Henry VIII.

Henry's matrimonial disputes led to a break with the church in Rome, and in 1536 he closed down the monasteries of England, throwing a lot of monks out of work. Many of these men were knowledgeable about brewing and set up business commercially.[14] Some attached themselves to a noble's household, and some went into business for themselves. A standard method of brewing evolved. Malted barley (dark and smoky, because it was kilned over open hardwood fires) would be mashed (steeped in water) and fermented at warm temperatures (read "ale yeast") to yield brown ale. This brew would be very strong, by today's standards, at about 13% abv. The malt would then be *remashed* (steeped again in water), producing a weaker beverage they called beer (9% abv), though it may not have been hopped. A third mashing of the same malted barley would yield an even weaker brew—about 5% abv—called "small beer."[15]

In the eighteenth century some London brewers began to depart from this practice and produced an odd, dark brew that was popular for two centuries: porter. This beer was made by mixing fresh brown ale with stale beer, i.e., beer that had been aged for some considerable time (perhaps a year) and had gone slightly sour. The resulting drink was malty, smoky, and with a sour, tangy aftertaste. There were several versions of this beverage in different parts of the country, and it evolved over time. The most important consequence for modern tipplers was an offshoot perfected by one Arthur Guinness of Dublin. His "stout porter" (see fig. 1.7)—later abbreviated to stout—thrived then as now.

With the industrial revolution came the pale ale revolution. In the late 1700s canals were built across Britain. Their main purpose was to

14. It is perhaps significant that Burton, one of the main brewing towns in England, evolved around an abbey.

15. Hence our derogatory expression for something weak or insignificant.

Figure 1.7. A pint of full-bodied stout, nutritious and with a long-lasting creamy head, is a meal in itself.

bring coal to the cities cheaply—land transport in the age before paved roads was slow, inefficient, and expensive. Cheap coal was necessary to drive the new-fangled steam engines of James Watt that were powering up the world's first industrialized economy. Both cheap coal and steam engines would greatly influence beer evolution. I will get to the contribution of steam engines after the intermission, below; here we will examine how inexpensive coal gave rise to pale ale.

Pale ale evolved in the eighteenth century. It was produced from pale (unsmoked) malt, kilned from coal or coke fires that did not darken the malt. In those days coal, and hence pale ale, was expensive. Consequently, the new pale brew appealed only to the upper echelons of society, such as the officer class in the British Army. The most important overseas posting for these army officers was in India, the "Jewel in the Crown" of the British Empire. An enterprising brewer decided to brew pale ale for Indian Army officers and send it out to India by ship.[16] To make sure that the beer did not go bad on the long outward journey, it was heavily hopped.[17] Hence India Pale Ale (IPA). Later, coal

16. Outward-bound shipping charges were low because trading ships brought goods from India and returned empty. So beer could be brewed in England and marketed in India at an affordable price.

17. Yes, I know, if it is hopped we should call it beer and not ale. IPA is the exception to the rule, in this book as elsewhere. The Tree Brewing Company of British Columbia, Canada, makes an excellent IPA—the hoppiest beer that I have found.

became inexpensive, and at the same time beer taxes were reduced or eliminated (to cut down on spirit drinking; see fig. 1.8). These changes combined to greatly reduce the cost of pale ale, making it affordable to the general public in Britain, and not just the well-heeled. The result was an explosion in pale ale sales, as you can see from the graph in figure 1.9. Pale ale was seen as progressive and modern, the beer of the railway age, and was drunk by the burgeoning middle class, who were benefiting most from the industrial revolution (see fig. 1.10). The older style of brown ales and beers were considered drinks for the workers.[18]

The industrial revolution generated glassware; tipplers were now drinking beer out of glasses rather than, as earlier, earthenware or pewter pots. Put pale ale into a glass, and your average tippler will naturally cast a keen eye on the liquid. To clarify the brew and make it appear more attractive, *finings* were added to beer in the barrel. Initially, these finings—which cause suspended solids to precipitate out, leaving a clear (*bright*) liquid—were *isinglass* (ground-up fish bladders[19]) or *Irish moss* (a type of dried seaweed). Nowadays, inexpensive gelatin performs the same function, though Irish moss is still used by some homebrewers.

During the 1800s, as the industrial revolution gained momentum, English beer production changed remarkably, and these changes were quickly adopted in other countries, as we will soon see. Here is a description of some of the main types of beer that belong to the English style.

18. Beer at this period was guzzled in enormous quantities by beer drinkers of all classes, in all the countries we are considering. Thus, for example, Benjamin Franklin recorded the drinking habits of workers in a London print house in the mid 1700s: each worker had a pint before breakfast, a pint between breakfast and dinner, a pint at dinner, a pint at 6 p.m. and a pint when he had finished work. Given the great strength of beer at that time, the quantities drunk are literally staggering.
19. The connection is not at all obvious ("How will I clarify my brew? I know, I'll add ground-up fish bladders!"). We have to thank William Murdoch for this efficacious discovery. He was a member of the Lunar Society—a kind of amateur "think tank" in late-eighteenth-century England—and an employee of James Watt. Murdoch also invented the sun-and-planet gears that were an important component of Watt's steam engines.

Figure 1.8. Social commentator William Hogarth on alcohol use and abuse—an early "health warning." In "Gin Lane" (a) Hogarth depicts the evil influence of strong spirit (cheap gin) in mid-eighteenth-century England. Gin was taxed much less than beer at the time. Contrast the wretchedness depicted in this engraving with Hogarth's vision of how things should be if taxation was more sensible and less punitive to brewers: in "Beer Street" (b) we see a happy, healthy, and thriving community.

(b)

Pale ale/IPA. Strong (originally about 8% abv), pale, and very hoppy beer. See figure 1.11 for a modern-day example.

Bitter. A pale ale variant, brewed with darker malts. Heavily hopped beer, from bronze to copper in color. Significant local variations.

Stout. Very dark, usually dry, and heavily hopped beer, with adjuncts such as flaked barley to add body. Chewy and nutritious. (See fig. 1.7.)

Porter. A stronger, though lighter-bodied and older version of stout.

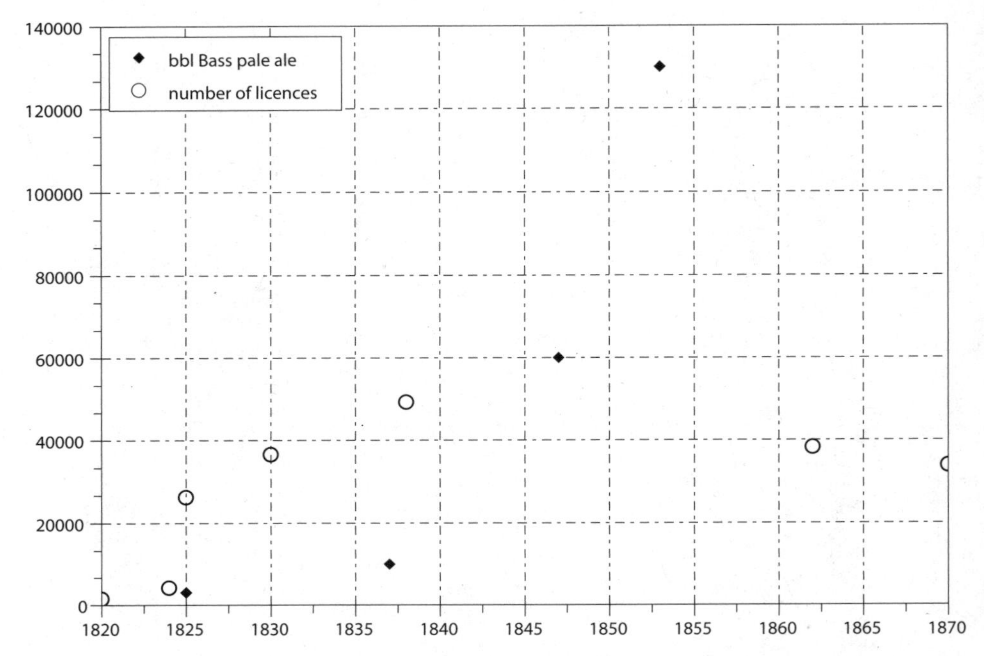

Figure 1.9. The number of brewing permits in Great Britain grew rapidly following a relaxation in the beer taxation laws, around 1825. The production of Bass pale ale also grew rapidly as a consequence of the industrial revolution. Bass pale ale was exported to mainland Europe, but within a few decades these exports would be eclipsed by the growing popularity of Pilsner beer from Germany.

Mild. Dark ale, lightly hopped, and so not very bitter. Also not very strong. It was traditionally served while fresh because it did not keep long. Cereal adjuncts add a rich, grainy character.

Belgian Beer

Belgian ale is much older than the country. It is more variable than the beers and ales of the other traditions and, to an outsider, much more difficult to categorize and classify. I don't know if Belgian brewers beamed down from Mars, but for whatever reason the beers of Belgium evolved according to different rules, whereas the beers of adjacent nations were straightforward reflections of the expanding Czech-German style. (There are excellent Dutch and Danish beers, for example, but they are clearly of the Czech-German type.)

Ale flourished in Belgium from Renaissance times. Its importance may be judged by the imposing edifice of the Maison des Brasseurs, on the Grand Place at the heart of Belgium's capital city, Brussels. With poor roads and no means of preserving perishable products in those days, each village had its own brewery, and so there evolved many and varied styles of ale. By 1900 there were 3,223 registered brewers in Belgium, varying from small cottage industries to large, modernized breweries. Reflecting the international trend of the twentieth century,

Figure 1.10. The impressionist Édouard Manet was probably not intending to capture the rise of pale ale in this painting, *The Café Concert*—the brew here is probably lager, the dominant brew in 1870s Paris, where this picture is set—but another of his paintings explicitly includes Bass pale ale, which was exported across Europe at this time.

this number was reduced to 2,013 by 1920. There are powerful economic reasons for the merging of smaller breweries into larger ones, as we will see after the Intermission, but in Belgium's case there was another historical reason: World War I got in the way. By 1946 the number of breweries was down to 755.

However numerous the breweries, there have always been hundreds and hundreds of beers, ales, and lagers in Belgium. They make use of both ale and lager yeast, as well as of wild yeast: the wort of *lambic* beers is exposed to the open air, in the brewery attics, and is fermented by wild yeast that blows in from the valley of the river Senne. Belgian brewers make use of many different adjuncts, particularly fruit. Here is a very brief description of some of the principal styles of beer to be found in Belgium.

Figure 1.11. A wonderful English pale ale: Summer Lightning is brewed in southern England but is popular all over the country. Thanks to Sophie Green of Hop Back Brewery for this image.

Figure 1.12. Very weak by Belgian standards, at 4.8% abv, Steendonk is a refreshing wheat beer. I thank Andreea, from Brussels (www.belgian-beers.eu), for this image.

White beer (*Witbier* or *biere blanche*). Unfiltered, cloudy beer made from oats and barley, as well as wheat. Typically with orange peel or coriander added during the mash. (See figure 1.12.)

Bottom-fermenting lager. The Belgian version of Pils: fresh, bitter, and hoppy.

Trappist. Strong, top-fermented beers brewed by monks. (See fig. 1.13.) Typically, candy sugar is used, and the beer is *bottle-conditioned* (i.e., yeast continues to ferment in the bottle, to provide carbonation).

Lambic. A flat (no frothy head) wheat beer spontaneously fermented from wild yeast. Matured in wooden barrels, it has a wide range of flavors.

Figure 1.13. At 9.2% abv, Rochefort is a strong Trappist beer. Thanks to Andreea for this image.

Gueuze. A blend of young and old (though not yet fully fermented) lambic. Sparkling and sharp; its fans think it is the champagne of beers.

Amber. A top-fermenting beer made with caramelized malt. (See figure 1.14.)

Fruit beer. Mixtures of fruit and lambic. For example, *Kriek* contains 1kg of cherries for every 5L of lambic (the equivalent of about 2 pounds per gallon).

INTERMISSION

Now is the time to get off your seat and stretch your legs, take a stroll to the fridge, and quaff a few beers, perhaps accompanied by beer nuts. I have reached a natural interlude in my summary of beer history, and you deserve a break. Instead of filling this time with banal adverts for hair shampoo and forthcoming reality TV shows, however, I propose to further improve your minds and morals by relating to you a poignant tragedy of Greek epic proportions. Afterwards we will see the relevance to the history of beer of this sad (yet strangely hilarious) drama.

The Courtship of Adipose Al: A Cautionary Fable

Aesop may have had moral instruction in mind when he wrote his famous fables—and they have survived for 2,500 years or so, which is good enough for me. Perhaps in AD 4500 a young tippler may read this tale and emerge from it a sadder but wiser man (or woman).

Adipose Al was a larger-than-life citizen of the city of Milwaukee, in the state of Wisconsin, in the United States of America. A single man of convivial disposition, Al chose to spend much

Figure 1.14. Vieux Temps ("The Good Old Days"), another light Belgian beer, this time an amber. Thanks once more to Andreea for this image.

of his spare time in Chuggaluggers, his favorite downtown bar, which served—what else?—the economic mainstay of old Milwaukee: fizzy, pallid lager beer that was produced and consumed on an oceanic scale. It was produced by giant brewing companies, packaged into colored cans with different names, and adorned with small pictorial illustrations suggesting a long brewing tradition from old Europe, and further suggesting many prizes won in ancient beer festivals that Al had never heard of. The evening before our story begins, Al was at Chuggaluggers, in his usual seat at the bar, quaffing Lyte beer (brewed about as well as it was spelled) at a rapid rate, as contestant no. 17 in a "Hunt the Flavor Molecule" competition. This contest was keenly fought, in friendly though raucously vocal style, every Friday night (perhaps that should be "Fryday nyte") at Chuggaluggers. The winner was the guzzler who claimed to have tasted the mythical flavor molecule that was rumored to lurk at the bottom of every tenth glass of Lyte. (Misanthropic cynics claimed that these rumors were a marketing ploy of the Big Brewers.)

This evening saw Adipose Al waddling towards Stunners, a Nyte Klub for the under-35s. Al was 45 years old, but he was not concerned about being denied entry to Stunners, because in this case 35 referred to IQ and not age. As he squeezed through the door and handed over his entrance ticket—a prize that he had won at Chuggaluggers—Al was not thinking of Lyte but of Love. This Saturday night would be different, and Al was going to score with a hot babe. He had showered (it was July, and Al showered every July, whether he needed to or not) and put on his best sneakers, track suit pants, Meatloaf T-shirt, and Milwaukee Brewers baseball cap (backwards, in deference to current fashion and to confirm that he qualified for entry into Stunners). He had taken on board some refreshment and advice at Chuggaluggers—a lot of refreshment and a little advice—and was now readying himself to approach Big Brenda and ask her to dance.

There she was at the bar. Brenda was a diminutive (compared with Al) 350 pounds, squeezed into a dress meant for a much more slender member of the fair sex, and Al was smitten. He flubbered towards her.[20] The light glinting off the sequins on her dress (which must have been held together by high-tensile steel thread) combined with the strobe disco lights and the July heat was making Al feel a little queasy. The gallon of Lyte he had taken on board didn't help, though Al considered that such refreshments made him more alluring to members of the opposite sex, notwithstanding his slurred speech and the inverted-V beer stains on his Meatloaf T-shirt, from beer that had missed his mouth in the frantic effort to get it down.

Our hefty hero had enough wit left in his befuddled brain to heed the advice given him by his friend Valentino, the "Bawdy Belgian." Valentino had a reputation as a lady's man among Al's coterie, for the simple reason that he enjoyed the company of young ladies much more often than they did. Suave, sophisticated, lithe and handsome, Valentino was that very evening entertaining Svelte Suzie at his bachelor pad. Valentino forswore Lyte and brewed his own beer. Valentino had tried on occasion to convert Al and his friends to the virtues of homebrewing ("Flavor molecules abound, my friends! Save your hard-earned money to spend on the ladies, and make beautiful beer that tickles your taste buds!"), but it was no use. Years of Lyte had caused Al's taste buds to wither, and the big-buck advertising of the Big Brewers had poisoned his feeble mind. For Suzie, Valentino had produced a well-aged Belgian Kriek beer, served in an elegant, slender glass with a cherry on top. Valentino murmured to Suzie about his continuing the age-old traditions valued by Trappist monks, but Suzie, who saw the way he was looking at her, doubted that. Even so, she did not mind his attentions.

Anyway, Valentino had suggested a few smooth words for Al to employ when asking Big Brenda to dance with him. Al, perhaps surprisingly, had remembered enough of these words and blurted them out in a sufficiently coherent manner for Brenda to under-

20. I have purloined the word *flubber* to describe the gait adopted by Al on this and most other occasions. Think of a bull elephant seal moving heavily across a sandy beach towards a much smaller female, and you will get the general idea.

stand. She tottered gingerly to her feet, and her beau led her to the dance floor. Amazed onlookers were later to report on what happened next, and accounts differ in the general mayhem and stampede of panicking dancers. One survivor (a physicist) noted the whirling gyrations of Al and Brenda, sweating profusely under the strobe lights, and thought unkindly that the attraction between them must be purely gravitational. Another survivor (a plumber) heard the first volcanic rumblings from Al's capacious beer belly, as the gassy Lyte began to churn . . .

A sense of delicacy and a consideration for my reader's finer feelings prevent me from describing in detail the final act of this tragedy. Suffice it to say that when Al exploded, the disco floor, walls, and ceiling were dappled with splashes of color that rivalled the gaudiest of disco light shows. Moralists like to point out the poetic distribution of Al's major organs. His heart and one other organ landed in Brenda's lap, his stomach in the restaurant, his kidneys in the rest room, and so forth. Most fittingly of all, perhaps, his gigantic bladder was donated to Chuggaluggers, and even today it reclines on his favorite barstool. Of course its contents were drained, chilled, and then recycled because, as every beer drinker knows, the sole difference between Lyte entering the body and Lyte leaving the body is temperature.

You see what I mean about epic proportions? Well, in fact Al's rotundity is irrelevant to the main lesson of my fable. His considerable weight was a literary device that I adopted to make his explosion seem more dramatic. The moral of the tale might have been clearer had I added the observation that Beanpole Bill of Buffalo exploded that same evening, all 6 feet 10 inches and 75 pounds of him, due to gas pressure from a gallon of Lyte. Thus, Al and Bill have taught us that (a) if you drink homebrew then you are lithe, handsome, and attractive to other people; but (b) if you drink Lyte beer then you are a no-brain who is doomed to detonate.

The fable may have given you the impression that I have a particular dislike of Milwaukee brewing companies. Not so: many other giant macrobreweries (and henceforth the word "macrobrewery" will be abbreviated *MB* in this book) all over the world, not just in the United

States, also make vast amounts of this pallid, ghostly imitation of Czech or German lager. In fact, the regular beer of most MBs is bland and characterless; their light beer is just an extreme version of it. Such beverages—regular and light MB beer—are popular, partly because of massive advertising but also partly because they are served cold. On a hot day, I find, almost any cold, fizzy drink is refreshing. In England, where American, Australian, and European MB beers have flooded the pubs over the last couple of decades, they are known derisively to true beer fans as *Eurolagers*. Wikipedia reports that in the United States there are two exactly equivalent words in current vogue among beer aficionados: *Budmilloors* and *Macroswill*.[21] I use the last of these three words in this book to describe MB beers. It is descriptive, it emphasizes that these bland brews are the product of MBs, and it does not suggest that any particular brewery or continent is solely responsible for inflicting tasteless beer—and in particular light beer—upon the peoples of the world.

It may have dawned upon a few readers that their author has not been wholly unbiased in his description of mass-produced MB beers. This is true and is the reason I have provided an intermission at this crucial point in the history of brewing. The facts of economic life and the industrialization of beer production have inexorably led over the last century to a contemporary beer world (the English-speaking part of it, at least) that is dominated by Macroswill. I have been able to maintain an unbiased view of beer history up to the intermission, but henceforth my preferences may be evident. As you are about to see, beer history from the late nineteenth century to the present day is one of industrialization, corporate mergers, globalization, and the rise of Macroswill. The response of beerophiles over the last third of a century is a natural and telling reaction. Now please take your seat, and I will resume the history lesson.

FROM MASS PRODUCTION TO MACROSWILL

Mass production of beer followed from the technological developments of the industrial revolution, initially in England in the early nineteenth century, and gathered pace as industrialization spread to

21. These words and others are discussed by Enkerli (see bibliography).

other countries. The high-pressure steam engine of James Watt started the ball rolling and greatly assisted the brewing process. Transporting raw materials to the brewery, and heating large quantities of wort, became less expensive. Saccharometers (which measure the sugar content of wort) and inexpensive thermometers were introduced in England to improve the control of beer production, resulting in less spoilage and more consistent brews. These innovations spread to Europe and North America.

Later in the nineteenth century, an understanding of the biochemical processes of brewing permitted a diversification of beer flavors and better control of the brewing process. Louis Pasteur, in France, finally understood that microorganisms called yeasts were responsible for fermentation. Emil Christian Hansen, a Dane, isolated different strains of yeast cells, thus enabling brewers to consistently brew beers of desired taste and character (recall that half the flavor compounds found in beer originate with the yeast). Then an Englishman, O'Sullivan, understood the role of enzymes in the biochemical process. The German von Linde introduced compressed gas refrigeration, as we have seen, thus enabling lager beer to be brewed in summer. Brewers were finally free of the seasonal dictates: they now knew why lager had always brewed better in winter (bottom-fermenting yeast prefers low temperatures), and they were now able to brew it in summer because of refrigeration. Another consequence of refrigeration, just as significant, was that beer distribution became much easier. Beer barrels could be transported great distances, and so the market for the product of a given brewery was geographically expanded. Sales increased, and the size of breweries increased as a result. This phenomenon occurred in all the industrialized beer-producing countries and nowhere more so than in the United States. So, to understand this expansion phase of beer history, I can do no better than look at the evolution of beer production in America.

Beer was part of the American colonies from the beginning. It is said that the Pilgrim Fathers landed farther north than they had planned, at Plymouth Rock, at least in part because they had run out of beer.[22]

22. "We could not take time for further search or consideration; our victuals being much spent, especially our beer" ("Bradford's and Winslow's Journal," in Stedman and Hutchinson, p. 125).

Unsurprisingly, in the early decades of the new Republic, it was the English style of beer that dominated; it was brewed locally and was also imported. In the seventeenth and eighteenth centuries, breweries in the United States (and in Canada) brewed brown ales, porter,[23] and pale ales. In the mid-1800s, following unrest in the old country, millions of Germans immigrated to America and brought their own tastes and skills to the American brewing scene. Many of the big U.S. brewers became established at this time: Busch, Pabst, Schlitz, Ruppert, and Ehret were all German immigrants, arriving at about the same time as the new bottom-fermenting lager yeasts, when the expansion of beer production through industrialization was getting under way. The result was a near-total domination of U.S. beer production by German-Americans brewing Czech-German–style lagers.

Cold beers (i.e., lagers) are naturally more popular in hot countries, and the weather gets much hotter in most parts of the United States in the summer than it does in England. Therefore, cellar-temperature English-style beer became less attractive than a cold lager to, say, an American worker just finishing his shift in a sweltering Chicago factory, or a settler in the expanding Southwest.[24] The combination of industrialized production and increased distribution, resulting from refrigeration and the expanding rail network, led to a phenomenal growth in the size and number of breweries, as you can see from the graph of figure 1.15. The late nineteenth century was the age of the beer barons in the United States. Anheuser-Busch pioneered the wide distribution of beer via refrigerated railcars and a network of ice-houses, as part of its goal of becoming the first brewer to offer a beer nationwide. Captain Pabst of Milwaukee became the first baron to sell one million barrels of beer in a single year. His products were sold in every major city in the United States. Pabst's great Milwaukee competitor Joseph Schlitz

23. George Washington was fond of English porter. Then in 1789 he switched to a porter brewed in Philadelphia, following his "buy American" policy.

24. There is a popular myth in North America that English beer is served "warm." This is not the case: "warm" is a relative term. English beer is served warmer than lager beers of the Czech-German style, at cellar temperature, which is a few degrees cooler than room temperature. In recent decades "cold" beers have also appeared in England (among English-style beers as well as lagers), perhaps due to warmer summers and aggressive marketing by MBs and the globalization of beer.

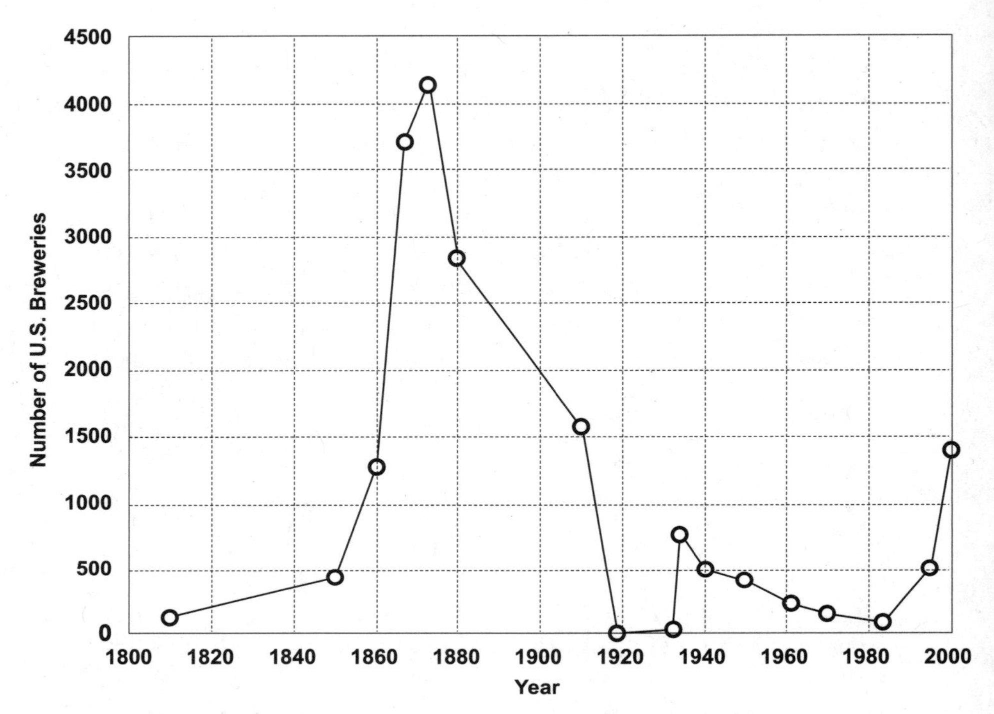

Figure 1.15. The number of breweries in the United States grew rapidly as the country industrialized during the third quarter of the nineteenth century. The trend since then has been of brewer conglomeration and merger. In the graph, zero breweries corresponds to the prohibition years, during which some breweries stayed in business by brewing "near beer" (very-low-alcohol beer)—but that doesn't count in this book. During the last 20 years of the twentieth century, the trend of falling brewery numbers was reversed due to the rise of microbreweries.

opened opulent beer gardens to attract customers. The beer barons made themselves known and their products popular by high-profile sponsorship (of baseball teams, for example), nationwide advertising, and efficient production (and so, affordable prices).

Increasingly, automation took over beer production. Automatic bottling (and later, canning) made for efficient distribution. Nowadays, of course, every aspect of the brewing process is computer-controlled, but even in the early 1900s automation helped drive down the price of producing a barrel of beer and increased the number of barrels produced. Beer was by this time a mass-production industry, not the cottage craft of a century earlier. In some ways this was good—quality control had improved greatly and so beer was less variable, and kept for longer. It was available in isolated regions far from the breweries. On the other hand, when the beer from single, large breweries became available nationally, small-scale local breweries were driven out of business, so that the varieties of beer tumbled. Often these local breweries had taken advantage of local conditions—soft water, say, or hard water, or good local hops—to create a tasty product that became part of the local tradition. Typically, across the United States (and England and Germany and Belgium and the Czech Republic, and everywhere that beer was produced) large brewers would take over small breweries, so that while the volume of beer increased, the number of breweries and the variety of beers fell. This trend began at the end of the nineteenth century and continues almost to the present day (see figs. 1.15 and 1.16).

Thus the twentieth century has seen an erosion of traditional beer

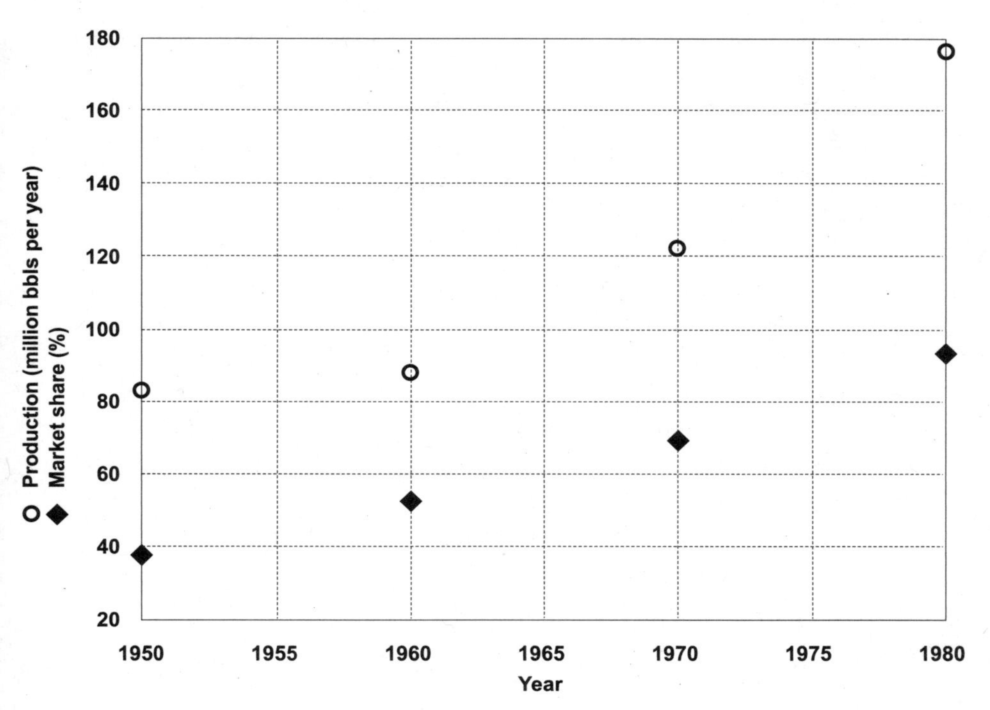

Figure 1.16. Production figures for the top 10 U.S. brewers, 1950–80. Note the almost total dominance of market share by these 10 brewers in 1980.

distinctions that were based upon origin, materials, and brewing methods. Here is an example of what I mean. In the last couple of decades the big MBs have switched from smaller, traditional fermentation vessels to large cylindro-conical tanks, for reasons of efficiency (separating the liquor from sedimented yeast). This change has, in most cases, obliged the brewer to change yeast variety. The traditional strains of yeast were selected to produce good beer in traditional fermentation bins, while the new fermenters needed different yeast qualities. Yeast is now selected for reasons of efficient production rather than beer flavor (again, half the flavor compounds in traditional beers come from the yeast). Stainless steel tanks are more hygienic than wooden vats but impart no flavor or character. In the 1960s metal kegs replaced wooden barrels. Kegs are less expensive, more durable, and more hygienic; they are easier to tap; and they simplify filling and cleaning. And, unlike wooden vats, they add zero flavor.[25]

So, you can see how Macroswill arose as a consequence of large-scale production. Take another example: beer maturation. Traditionally, fermentable sugars or *krausen*[26] would be added to the green beer in the barrel. Residual yeast in the beer would work to create carbon dioxide, which carbonated the beer, generating pressure and purging the beer of undesirable volatile compounds, which would be vented when the barrel was tapped at the point of sale—the pub. The beer would be served by bar staff who understood about their beer and would serve it only when it had "conditioned." Nowadays, it is judged to be more hygienic (and less expensive) to carbonate the beer artificially. Instead of allowing the beer to condition naturally, which took a lot of time (weeks, in the case of English beers and ales, and months, in the case of Czech-German lagers), manufacturers accelerated the process artificially by adding tannins, adsorbents, or enzymes to degrade the unwanted compounds (such as unfermented dextrins) much more quickly. Of course, without true maturation the flavor and character of beers is altered for the worse.

25. The extra flavor does not always come from the wood, as for whiskey when it matures in barrels, because beer barrels are often pitched (see chapter 6). Barrels are more porous than kegs, and beer in a barrel ages and develops more character than it does in a sterile keg (and it goes sour more quickly).
26. Fermenting wort. The German beer purity laws forbade adding sugar, and so their lagers were conditioned with wort.

Today MB beers are brewed to a high gravity, filtered, and then diluted to the desired strength with deoxygenated, carbonated water, before being pasteurized (in some countries) in the bottle or can, or in the metal kegs, by heating to 70°C for 15 to 30 seconds. All of these processes improve hygiene and reduce costs, but they also reduce the flavor of the beer. This sanitized Macroswill can be stored for long periods and distributed across the world. As long as it is fizzy and is served cold, and is promoted by TV advertising, it sells.

REAL ALE AND MICROBREWERIES

I hope that it is unnecessary to point out that some mass-produced beer is good stuff: not all MB beer is Macroswill. For example, Guinness is a mass-produced beer with flavor, body, and character; it sells in 120 countries and has totally dominated the stout market for decades. Domestic German beer is mass-produced, but of very good quality. (See fig. 1.17.) The trend, however, of larger and fewer brewers making more and more widely distributed beer is clear. Bland beer is cheaper to make (if you produce millions of barrels, even very small production savings will accumulate significantly), and it appeals to a mass market, in ways that local brews (geared for quixotic local tastes) cannot.[27]

On the other hand, over the last third of a century it seems that a discerning section of the population, in several countries, has decided that bland Macroswill is just too bland. To save money, many MB beers partially substitute rice for malted barley; rice contains lots of starch, which ferments perfectly well but imparts little flavor. Light beers have lower carbohydrate levels (and, as a consequence, reduced body and flavor), but most of the calories in beer come from the alcohol, so light beers are watered down more than regular beers. Thus, a typical light beer is 4.1% abv, whereas a regular beer is about 5% abv. Given the

27. Here is an analogy, possibly misleading. The interiors of new houses tend to be painted in a uniform bland pale color, to appeal to the greatest number of potential buyers. Nothing offends, except perhaps the blandness. Paint the same houses with strong colors and they will appeal more to a few people, and less to most others (because tastes differ). Macroswill has enough beer qualities to be recognized as beer (it is cold, fizzy, and alcoholic), but is bland. More strongly flavored local brews may not have a wider appeal.

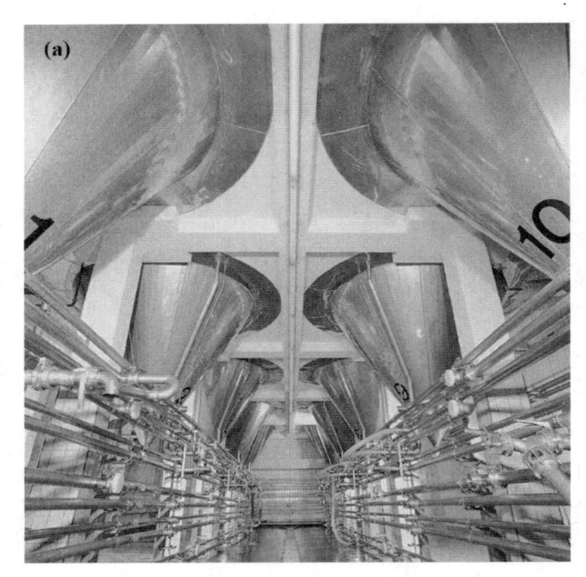

Figure 1.17. Large-scale beer production, German style. Photos courtesy of the Bavarian Brewers Federation, Munich, Germany.

blandness of most regular MB beer, what can we expect of the diluted light version?

In the early 1970s in England, a consumer lobby organization called CAMRA emerged. The *CAM*paign for *R*eal *A*le grew from 4 people to the present 84,000 members: beerophiles who were dissatisfied with Macroswill and were determined to do something about it. CAMRA publishes a Good Beer Guide, encourages small brewers who make traditional beer, and points out bad practices in the brewing industry. From the beginning, CAMRA's call for quality beer has resonated with many beer drinkers, and the number of small breweries that catered to this discerning market started to grow, reversing the 100-year trend of reduced brewery numbers and increased brewery size. In the late 1970s and early 1980s the same reaction arose in the United States. The first microbrewery (*μb*, in this book; the word itself dates from 1982) began operation in Sonoma, California, in 1977, and the first brewpub opened in Yakima, Washington, in 1982. A brewpub is a μb that serves food as well as beer. (For simplicity I will refer to both breweries and brewpubs as μbs.) As in Britain, these little μbs grew out of consumer pressure. Today there is a large and vociferous body of increasingly sophisticated American beer lovers who organize beer festivals, lobby the brewing industry, and energetically promote the brewing and appreciation of good-quality beer, ale, and lager.

A μb today is characterized as a brewery that produces fewer than 25,000 bbls of beer (or perhaps 15,000 bbls—it depends on whom you consult) per year. To put this in context, in 1966 Budweiser became the first MB brand to sell 10 *million* bbls in one year. The μb market is small, at about 2.5% of total U.S. annual beer sales,[28] but you can see from the graph of figure 1.15 how the number of breweries has grown since 1980—there are currently about 1,400 μbs. Some μbs have outgrown their 25,000 bbl limit. (The Sierra Nevada Brewery of Chico, California, was the first to do so. Sam Adams Brewery, of Boston, is currently the nation's largest μb.) Some of the older regional brewers have reposi-

28. But almost 10% in Oregon. Interestingly, the distribution of the so-called "craft beer" revival is far from uniform, being most heavily concentrated on the coasts: in the Northeast (35%) and along the Pacific coast (32%).

tioned themselves in the marketplace and now produce good traditional beer (see fig. 1.18).[29]

Whatever the brewery, whatever the country, the consumer reaction to Macroswill has generated beers that are more traditional in conception: they typically are made in small, local breweries from malted barley (or wheat or . . .) and plenty of hops, but with little added sugar, rice, or other fermentable materials that contribute negligibly to flavor. These beers are often unfiltered and unpasteurized, since filtering and pasteurization remove flavor. Note the emphasis on beer flavor and quality (at the price of increased production and distribution costs).

There are many beer Web sites that opine about all aspects of beer. Some of these Web sites express the ventings of a few conspiracy buffs, who see dark forces at work in the MBs' response to μb growth. Some MBs reacted to the growth of μbs by producing new lines of good beer, in small quantities and available at only a few of their outlets. Initially, this response was considered a good thing (though perhaps half-hearted and brief)—MBs were being educated by public opinion, and the standard of beer was being raised to earlier levels. Some MBs preferred to offer their "real ale" or "traditional quality beer" anonymously—i.e., they marketed them under a different name. Conspiracy theorists see these "stealth brews" as Trojan horses, sent in by the MBs to undermine μb sales and prestige. Thus, the theory goes, if the stealth beers sell well, then the MBs are pleased because they are cutting into the μb market. If the stealth beers sell badly, then the MBs are pleased because they have tarnished the name of "real ale" and "traditional quality beer." I would like to think that this response is too cynical.

The growth of traditional small breweries over the last 30 years has taken several forms: craft regional breweries, microbreweries, brewpubs, U-brew outlets, and homebrewers. The current umbrella term for the multitude of beers, ales, and lagers that are brewed by these connoisseurs is *craft beers*. The Brewers' Association "Beertown" Web site defines craft beers as follows: "Craft beers are produced with 100% barley or

29. I have in mind here, particularly, the Anchor Brewing Company of San Francisco. This company's "steam beer" is produced in very shallow fermentation vessels using bottom-fermenting yeasts working at a temperature normally associated with top-fermenting ale yeast.

Figure 1.18. Anchor Steam Beer from San Francisco. Another unusual combination of top-fermenting and bottom-fermenting traditions, this excellent beer is a very popular regional brew.

wheat malt or use other fermentable ingredients that enhance (rather than lighten) flavor. Craft beers only come from craft brewers." A "craft brewer" is characterized as "small, independent and traditional." The consistent theme here, across countries and threading through the last three decades, is a reaction against large MBs and a concern for reviving local, high-quality cask-conditioned brews. This theme may seem like the only common factor in a diverse and rapidly changing sector of the brewing world.[30] A few of the most vociferous advocates of the beer revival movement verge on paranoia,[31] but most craft brewers are united by a desire to make (yes, and to drink) good beer the way it used to be made.

30. Many μbs fold, to be replaced by new ones, and many homebrew stores and U-brew facilities come and go. Additionally, the U.S. national beer culture is diversifying (in large part due to the craft-brew influence) from the historical beer-drinking types: blue-collar males or college students or mainstream sports enthusiasts.

31. I have parodied the sense of persecution (Big Brother, in the guise of unthinking, uncaring all-powerful MBs, is out to get us) in my fable. I hope that most of you recognize this fact, and do not think that you will actually explode by drinking a gallon of Lyte. Hmm, on the other hand . . .

The next chapter of this book is concerned with the activities and products of the smallest-scale craft brewers: the homebrewing fraternity. Homebrewing makes μbs look big: perhaps I should refer to a homebrewer as a nanobrewer (*nb*), since the annual production of a typical homebrewer is at least a thousand times smaller than that of a μb. Homebrewing arose as part of the craft beer revival, and also as a reaction to the increasing cost of commercially brewed beer (even the best-quality homebrewed beer costs under a dollar a liter to make). For various reasons, homebrewing was an illegal activity in many countries until a few decades ago[32] although, one suspects, it has always been a factor of domestic life. It is to homebrewing that we now turn.

32. Homebrewing without a license and without taxes became legal in Great Britain in 1963, in Australia in 1972, and in the United States in 1979.

Two

How to Make Good Beer at Home

You can't be a real country unless you have a beer and an airline—
it helps if you have some kind of a football team, or some nuclear
weapons, but at the very least you need a beer.

—Frank Zappa (1940–1993)

Once, during Prohibition, I was forced to live for days
on nothing but food and water.

—W. C. Fields (1880–1946)

HOMEBREWING AS AN EXERCISE IN PHYSICS

Brewing is a rewarding exercise not only in fun but also in physics. In this chapter I get to tell you how I brew my own beer. Mine is a simplified, pared-down, minimal-effort method of performing a full-mash brew and produces good beer. (Pardon my boastfulness, but this is not the time for modesty.) At least, my friends who have sampled it over the years say the beer is good, and most of them are experienced experts on the subject of sampling beer. In the context of this book, sharing my brewing experience with you also serves to highlight the role that physics plays in the process. As a physicist, it has always struck me while brewing that there is a lot of physics on show at all stages. I will highlight the type of physics that we see as we go along, since several facets of brewing physics are discussed in later chapters. Thus, for example, when we pitch the yeast we will encounter population dynamics, as the yeast cells reproduce and adjust to the changing environment within the wort. Yeast population dynamics is the subject of chapter 3. Similarly, the mashing and boiling stages of homebrewing involve thermodynamics; this topic, as it pertains to brewing, is discussed in chapter 4. When fermentation is well under way, bubbles are produced; we will examine beer bubbles in chapter 5.

So, in the pages to follow, you will learn not only how to make beer but also where physics imposes itself upon the process. Analysis of these physical topics will reveal, for your edification and enjoyment, how physicists can apply their trade to an everyday human activity such as brewing, and not just to exotic, remote subjects such as cosmology or quantum mechanics. And afterwards you get to sample the fruits of our labors. Cheers.

EQUIPMENT

The first, vital piece of equipment that you need[1] in order to satisfactorily brew beer at home is a home. Sadly, these places are often not compatible with successful brewing, for sociological as well as pragmatic reasons. You may share your home with a wife,[2] and she may object to the kitchen getting steamed up and the whole house smelling like a brewery (which it is, though in my experience it does not help to point out this fact). You may have kids, in which case you need to carefully separate them from the brewing process, for the well-being of both. You may not have a room which can be kept at a reasonably constant and cool temperature—necessary for your beer to mature, as you will see. You may not have the space necessary for storage of equipment, or a laundry tub for all the cleaning of said equipment.

But let's assume that you live in a place which possesses an ambience that is compatible with brewing beer. What other equipment do you need? The following basic list of essentials can be purchased for a couple of hundred dollars at your local homebrew center:

1. Yes, you will be doing all the brewing. I will stand in the background and will occasionally offer a helping hand.
2. Many homebrewers do have wives, but I must admit that I have yet to meet a brewer who has a husband. In other words, homebrewing appears to be a male hobby. I will not go into the sociology underlying this thorny topic, for the very good reason that I don't understand it. The only point that needs to be made is that I am not trying to be sexist here, merely accurate.

- Plastic jugs (1 gallon is a useful size)
- Plastic fermentation bucket (6-gallon), with a faucet
- Plastic tubing, and connectors, for siphoning
- Air locks and rubber bungs
- A heating vessel (6-gallon plastic or metal kettle with heating element and a faucet)
- Thermometer and hydrometer[3]
- Glass or plastic carboy (6-gallon)
- Glass or plastic bottles, with airtight caps

Other bits and pieces will suggest themselves as you progress, but this list covers the basics. In addition to equipment, you will need supplies:

- Hops (type and amount depends upon the brew)
- Grain (pale malt), plus adjuncts such as flaked barley, crystal malt, black malt, liquid malt extract, corn, malted wheat, oatmeal, etc. (type and amount of adjuncts depends upon the brew)
- Sugar (corn or cane)
- Sachets of brewer's yeast

Again, these are the basics; you will find extra stuff to add to this list as you progress. The equipment is fairly self-explanatory, but the supplies call for more details. Hops are sold either as dried flowers or as pellets; both are acceptable. The type of hops that you use, and the amount, greatly influence the beer flavor. A good homebrew store will have dozens of different types of hops to choose from, and these will be labeled by their *alpha acid* content. You don't need to know what alpha acids are, but the higher the content in your hops, the more bitter your beer will be. Homebrewers will often buy a bittering hop with high alpha acid content and an aromatic hop with lower acid content. Two or three types of hop may be combined for a specific beer recipe.

3. The hydrometer floats in the wort and measures wort density, and is a useful indicator at several stages of the brewing process, as we will see. Some thermometers also float in the wort, but I prefer the more familiar bulb thermometer. It does not have to be very accurate but does need to be robust.

Here we will be making top-fermenting English-style beer, rather than bottom-fermenting lager, and so you choose a standard beer hop, Goldings. Lager beers have their own specialized hops (such as Saaz). To keep it simple, you decide to use the same variety of hops for both bittering and aroma, and so a general-purpose variety such as Goldings is a safe choice. Our brew will use 4 ounces of Goldings hops for bittering and 1 ounce for aroma. (If you don't like bitter beer, use less than 4 ounces for bittering.)

Pale malted barley is the basic fermentable ingredient of full-mash beer. We scorn MB staples such as rice and limit the amount of sugar in a brew. This is because grain adds flavor whereas rice and sugar do not. Pale malt is lightly malted barley. (There are several types with slightly different characteristics; you will find your favorite, and the staff at your local homebrew center will have plenty of advice.) For a 6-gallon brew I use 7–10 pounds of grain; let us say that you choose 8 pounds. This is more than some homebrewers use (and a whole lot more than MBs or even U-brewers use for the same volume of beer). They make up the difference with sugar, which boosts alcohol content without adding much flavor. Pale malt is typically 62% fermentable, which is to say that 62% of the grain weight is starch that will be converted by the yeast into water, carbon dioxide gas, and alcohol. Sugar is 100% fermentable.

You may want to consider two other variations of malted barley as adjuncts. Crystal malt—grain that has been heated longer than pale malt—is darker and imparts more color and a slight caramel flavor to the beer. Chocolate and black malts, so named for their colors, are barley grains that have been heated still more. They add color, a smoky flavor, and a grainy texture to the beer. I do not intend to describe beer recipes—many other books will do that—so will not suggest how much, if any, of these grains or other adjuncts to add. Half a pound of crystal malt, for body, and 2 ounces of black malt for coloring, you say? Good choice. So, the basic grist we will be using contains 8½ pounds of malted barley (the 2 ounces of black malt doesn't count because it is so toasted that most of the starch is unfermentable). Before leaving the subject of grain, I should point out that if, like me, you buy large 50-pound sacks of pale malt, it is essential that you have somewhere cool (but not cold) and dry to store it. If kept dry, malted barley grain can be stored for many months, or years.

Sugar is a complicated subject. There are many types of sugar, and not all of them are suitable for brewing. Your local brew store will have the right type. Corn sugar or invert cane sugar work fine, and I add between zero and 1 pound per 6-gallon batch of beer. Many recipes (particularly North American recipes) add much more sugar, and less grain. A small amount of flavor is imparted by brown sugar. I will discuss sugar in a little more detail in chapter 4. For this brew you decide to include 1 pound of corn sugar for your brew.

Yeast is also a complicated subject, with many brewer's yeast varieties on the market. You will almost certainly be restricted to buying small 5–7 g (less than ¼ oz) sachets described as suitable for a 6-gallon brew. This amount is not nearly enough. I always use at least two sachets; the reasons will emerge from our study of yeast population dynamics in chapter 3. Modern brewer's yeast is robust stuff, but even so, I tend to refrigerate the yeast sachets if they will not be used for some time, and then bring them up to room temperature a day before brewing. Tough little fellows they may be, but apparently they can suffer from thermal shock when the temperature is changed suddenly. Most of them will be laying down their lives for our beer, and so the least we can do is make life comfortable while they are with us. As mentioned earlier, yeast contributes half the beer flavor, and so you will have to experiment with different brands to see which works best for you. We will choose for our brew two sachets of top-fermenting ale/beer yeast.[4]

4. As we will see in chapter 3, yeast adapts to the wort environment in which it finds itself. So the yeast that survives one brew may be used again for your next brew, if stored in a cool place and covered carefully to avoid contamination. Such yeast will get going more quickly than sachet yeast, when pitched into the wort, because it has already adjusted to the wort conditions (assuming that your second brew is of similar type to the first). Even better is to use yeast that you have extracted from "bottle-conditioned" beer, which can still occasionally be bought from stores. This beer matures in the bottle and has a yeast deposit that must be carefully separated from the beer when pouring. Don't throw away the gunk—feed it with yeast nutrient (one of those extra items that are in all brew stores) and pitch it into your next brew. It will go like a rocket.

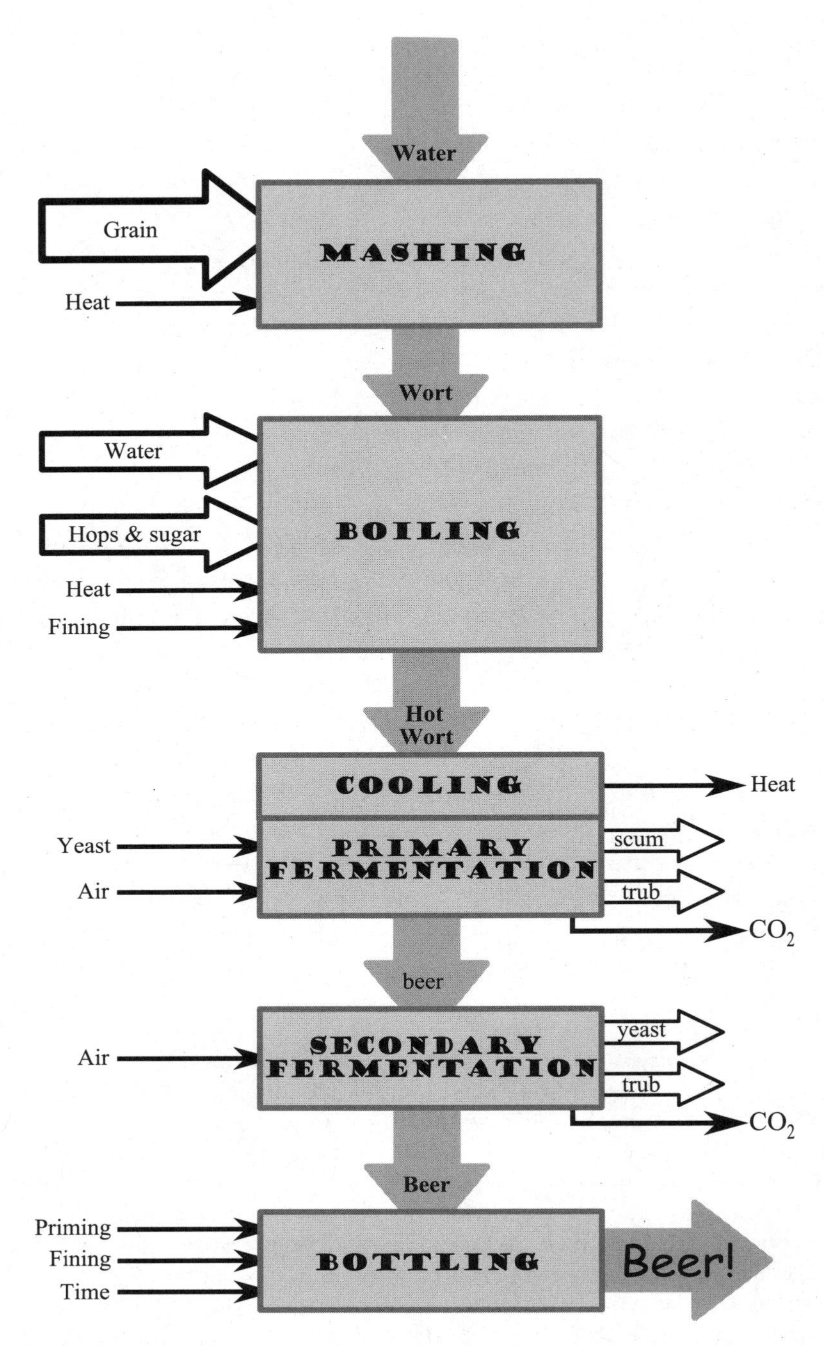

Figure 2.1. A home brewer's flowchart, showing the main steps in brewing and the ingredients and by-products.

You now have the basic ingredients for the brew that you are about to make. For future brews, no doubt, you will add different adjuncts to produce another combination of beer flavor, color, and body that you like. How do we combine these ingredients to make beer? The process of homebrewing, and the flow of materials, is shown in the chart of figure 2.1. The rest of this chapter is devoted to explaining the different steps shown in the figure.

MASHING

So now you have your grain—8½ pounds of pale and crystal malt plus a little black malt. To turn these grains of barley (see fig. 2.2) into wort requires cracking open the grains to expose the starch and then steeping them in warm water. Readers in England do not have to crack the grain that they buy: it arrives at the homebrew shop already cracked. The rest of us have to do it ourselves. It is a tedious and slow process if done by hand; you may want to indulge yourself and purloin an old electric food mixer. The result is that the grains are split, and you can see the difference between cracked and uncracked grain in figure 2.3. Cracked grain shows the white starchy interior, and each grain fragment is smaller than an uncracked grain.[5] The exposed surfaces of starch can now be dissolved in water.

The *infusion* method of mashing is traditional in English-style beers and among homebrewers. To infuse properly, it is necessary to add water to the cracked grains and heat it up gradually, holding the mixture (the *mash*) within a narrow temperature range for a while, and then further heating, holding, heating, etc. Thus, for example, one expert American homebrewer recommends the following pattern:

- raising the mash temperature to 95°F–100°F (35°C–38°C) and holding it there for 30 minutes;
- then raising it to 113°F (45°C) and holding for 1 hour;

5. Despite being smaller, the 8½ pounds of cracked grain occupies a larger volume (perhaps 5% larger) than the same grain before it was cracked. I would have expected the cracked grain to occupy a smaller volume, since the gaps between grains could be filled in, but this is not the case. Explanations from readers who have expertise in the field of granular physics would be appreciated.

Figure 2.2. Beer, growing in the field. Today barley is cultivated almost entirely for the brewing industry. Photo courtesy of the Bavarian Brewers Federation, Munich, Germany.

- then raising to 135°F (57°C) for 20–30 minutes;
- then raising to 150°F (65°C) for 45–60 minutes; and finally
- raising to 170°F (77°C) for 10 minutes.

Fear not: you can make perfectly good beer without being quite so precise. All this raising and holding makes the mashing process sound like a poker game, and I suspect that it is a bit of a gamble. For reasons that I will discuss more fully in chapter 4, the temperature throughout the mash is not the same, at any given time, and so for most homebrewers the precise scheduling of temperatures just outlined is not achievable. Read on, for a simpler and easier method that works.[6]

Fill your fermentation bin with 3–4 gallons of warm water (at around 150°F) and add the cracked grain. (Also add any other fermentable grain products that your recipe calls for, at this stage: flaked barley, malt extract, malted wheat, oatmeal, etc.) It is convenient to put the grain into a perforated grain bag first; this enables the water to seep through

6. It seems that the purpose of temperature-step infusion mashing is to reduce the possibility of haze in the finished beer. I find that my beer is clear (*bright*, in the language of brewers), in general, without the detailed temperature stepping. Anyway, as stated in the text, holding the mash at a precise temperature for a precise time is very difficult for the homebrewer.

Figure 2.3. Pale malted barley grains as supplied (*left*) and after cracking (*right*).

and get at the starch, while retaining the solid grain husks. Your thermometer will tell you that the mash (water plus grain) is now at a temperature of around 100°F–120°F. Go away and do something else for half an hour or so (I find that there is always plenty of cleaning and tidying up to do), then boil some water—enough to make up a total of 6 gallons—and add it to the mash.[7] The temperature should be around 150°F–155°F, and you can now go away for 1–1½ hours and clean some more. During this period the mash temperature should stay in the

7. Beer is 90–95% water (and more than that in weak beer), as is made clear in this old joke. Bartender, looking out of pub window: "It looks like rain." Grumbling tippler, holding up pint of beer and staring critically into it: "Yes, with just a touch of hops." Well, you get the point, I hope. In fact the type of water (hard or soft) makes a difference to the beer, and breweries were traditionally sited near sources of the best water type for their beer. Nowadays, the salts that are contained in water can be adjusted artificially, and some homebrewers do indeed add salts, which are readily obtained at any homebrew store. I have never bothered with this, nor do I use spring water. Good, clean water from my kitchen faucet works well enough, chlorinated though it is. (The water is about to be boiled, which helps.)

region 145°F–155°F (62°C–69°C). Because the temperature varies throughout the mash, however (a fact that you can easily verify by inserting your thermometer at different places), it is impossible to fix a constant uniform temperature. Consequently, I aim for the middle of the allowed range, and that seems to work. Above 155°F some of the enzymes produced by the malt are destroyed by the heat; below 145°F these same enzymes act sluggishly. In their comfort zone these enzymes convert the starch to fermentable sugars, and the purpose of mashing is to get as much of these fermentable sugars as possible out of the grain and into the water. So, maintaining mash temperature at about 150°F for an hour and a half does the job.

Of course, the mash will cool once you have reached the desired temperature, but over a period of 90 minutes it will not fall out of the acceptable range. At the end of that time (say 2 hours from the start of the mashing process) you remove the grain bag,[8] leaving behind a sticky solution of warm water that contains a lot of dissolved fermentable sugars: this is the wort. Separation of the grain and wort is often achieved by opening a tap at the bottom of the fermentation bin, so that the wort drains into your heating vessel, leaving the spent grain behind. One final trick that is used by all brewers—whether professional or amateur, large-scale or small—to extract the last few ounces of sugar from the mash is to add hot water to the spent mash, let it steep for a few minutes, then drain it off, adding the liquid to your stock of wort. This process is known as *sparging* (or *lautering*) the mash; sparging as I have described it is sometimes called *remashing*. An efficient mash process will remove as much as 90% of the fermentable sugar from the grain. In practice, homebrewers achieve rather less. You will know soon enough how efficient or inefficient your mashing has been.

BOILING

The next stage is simplicity itself—no need to measure temperatures here. Simply boil the warm wort for an hour. If your heating vessel is

8. Commercial brewers sell the spent grain (*draff*) to farmers as cattle feed—cows love the stuff. I operate on a smaller scale and give my spent grain to a neighbor for her chickens.

supplied with a typical 2- to 3-kW heating element, you will find that it takes about a half hour for the wort to come to a boil. Start your clock when the wort begins to boil. Maintain a good, vigorous, rolling boil—not a half-hearted simmer. The boiling serves several useful functions: it kills off any bugs that may be in your wort,[9] it extracts bittering acids from the hops, and it drives away undesirable protein compounds in the wort. This last function takes at least an hour to break up the proteins so that they form gooey clumps called *trub* that drop out of the wort once the boiling is stopped. So, boiling must last an hour or more; two hours won't do any harm if you have the patience.

When the wort is coming to a boil, it generates a sticky scum on the surface. This scum can overflow and cause a mess, so stand by during the last few minutes before boiling begins, and stir the scum back into the wort with a spoon if it looks as if it is going to make a bid for freedom. More on the sticky scum and the thermodynamics of the boiling phase in chapter 4, but here I will concentrate upon the practicalities of the homebrewing process, and move on to other matters. As soon as the wort comes to a boil, add your mixture of bittering hops. The hop flowers (see fig. 2.4) or pellets are usually added in a perforated nylon bag, which retains the hops while permitting the boiling wort to seep inside and extract bittering acids. If you are adding aromatic hops, pitch them into the wort 10 or 15 minutes before the end of the boil. You are adding a pound of sugar: pitch that in a few minutes before the end. Also, at this stage you may throw in a little Irish moss to fine the wort (it assists with precipitating out some of the gunk, but is not essential, in my experience).

Why boil the sugar? To kill any bugs that may be in it and to caramelize some of it for flavor. Why add the sugar only near the end of the boil? Because the heating element can get gummed up otherwise and is a real pain to clean.[10] More importantly, a gummed-up heating element

9. For this reason, we do not need to be particularly fussy about cleanliness before the boiling process. After the boil, however, we must always be aware of the possibility of contamination by environmental bacteria, and so must be careful to maintain hygiene.

10. You will clean the heating element after every boil as a matter of course if you have any sense, but this cleaning becomes much more difficult if caramelized sugar is stuck onto the element.

Figure 2.4. Hop flowers. Photo courtesy of the Bavarian Brewers Federation, Munich, Germany.

does not transfer heat efficiently, and this can cause the boiling to cease. The only recourse you have when this happens is to drain the hot wort into another vessel, dismantle the heating vessel and clean the element, put the heating vessel back together, reintroduce the wort and begin the boiling anew.

(I hope that this gumming up of your heating element does not happen, but if it does you may notice an interesting phenomenon that I am at a loss to explain. You will notice that boiling stops, which is the first indication that your element is gummed up. Your heart sinks, but then you observe something odd: concentric circles forming on the surface of the wort—standing waves. Placing your hand against the side of the heating vessel reveals the reason: the vessel is vibrating. Why? It must be the heating element that is vibrating, but why it should do so is a mystery to me. I realize that the electrical power supplied to the element must go somewhere. It usually heats up the wort and maintains a vigorous boil. It cannot do so when the element is gummed up, because the goo acts as insulation, and so the power must dissipate in some other way, but why does it appear as a vibration? Answers on a postcard please.)

Once the boil is finished, we sparge the hops. That is to say, we fish out the bag(s) of hops, place these in a jug, and add mash-temperature water. After a few minutes, we squeeze out the bags and add the water, now turned a pale yellow-green with hop oils, to the boiled wort. Once this is done and the wort is beginning to cool, we must be careful to avoid contamination by bacteria. The wort is full of fermentable sugars that are ideal food for brewer's yeast but also for bacteria; these bacteria will happily turn your wort into vinegar, or worse. So, once the boiling has stopped and the sparged hop water has been added, you drain the wort into a fermentation vessel and put on the lid. This will keep the wort safe while it cools. Many homebrewing books will advise you to cool the wort as quickly as you can—they will mention hot and cold breaks[11] and will cite the risk of bacterial contamination—but you have bought *this* book and so you just let the wort cool naturally. Some details of how you can accelerate the cooling process are provided in chapter 4, for those who feel the need to do so or are not prepared to wait several hours for nature to take its course. My experience is that natural cooling has no adverse influences upon the beer so long as a tight lid separates the wort from the outside world. So, let's take it as done. The grain mixture has been mashed to form wort, and the wort has been boiled and allowed to cool. What next?

PITCHING THE YEAST AND PRIMARY FERMENTATION

Your yeast sachets have been brought out of the fridge and allowed to warm up gently to room temperature. You have placed your hand against the side of the fermentation bucket, containing your lovingly prepared wort, and noted that it has cooled to body temperature. Now it is time to introduce the yeast to the wort, stand back, and let a miracle of nature take its course.

Well, you don't stand back just yet. The first thing you do is sterilize a thermometer and a brewer's hydrometer by placing them in a jug of just-boiled water for a few minutes. Take the lid off the fermentation bucket and check the wort temperature. If it is below about 100°F

11. Hot breaks are discussed in chapter 4. Cold breaks arise when, for example, you put beer into the fridge. Unfiltered beer can develop a *chill haze* that clouds the beer slightly. This happens sometimes with my beer, but not often, and not much. It does not detract from the visual appeal of the beer, and because it is not filtered, my beer retains a lot of its flavor even when cold.

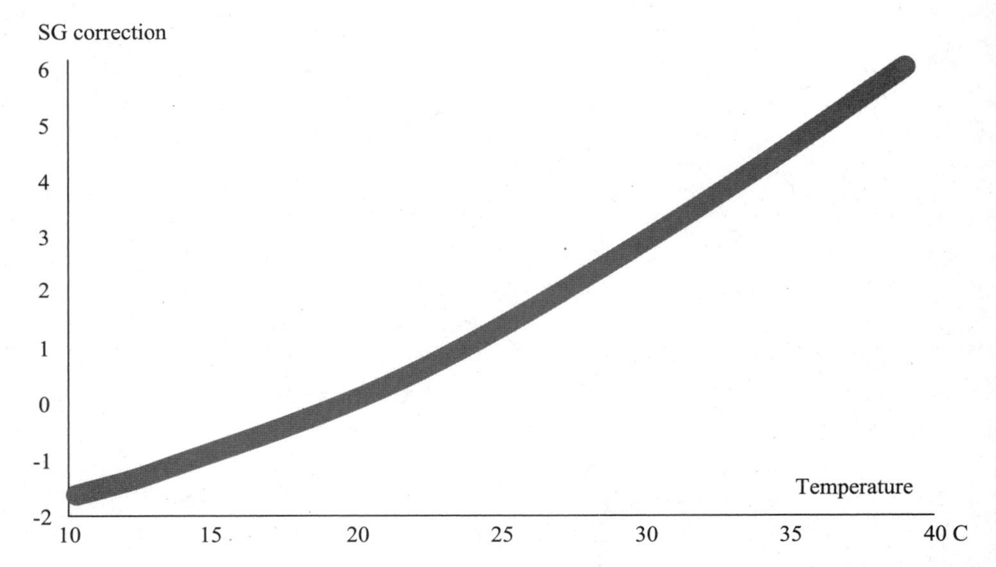

Figure 2.5. Correction to hydrometer reading of original gravity. My hydrometer assumes that the wort temperature is 20°C; if it is at a different temperature, then the corrections shown here must be added to the reading.

(38°C)—say body temperature—then that is cool enough for pitching the yeast. Any higher temperature will kill the yeast.[12] Now place the hydrometer in the wort and measure the *original gravity* (OG, which is the density multiplied by 1,000). You have used 8½ pounds of grain and 1 pound of sugar; if your mashing is about as efficient as mine then you will obtain an OG of around 1035–1040. That is to say, the density of the wort is between 1.035 and 1.040 times the density of water. The hydrometer measurements must be corrected for wort temperature; the amount of correction is shown in figure 2.5.

12. Often, homebrew sources will say that the yeast should not be pitched until the wort temperature is much lower—say 75°F–80°F—and it is certainly OK to wait until the wort has cooled to this temperature. The yeast that I use seems to tolerate body temperature (98°F, or 37°C), however, and anyway, the wort is cooling all the time. The steady fermentation temperature should be within a few degrees of 68°F (20°C), and this can be maintained by controlling the ambient temperature of the room in which you brew, or by adding supplementary heating (e.g., via a heating belt or pad) as required. The high pitching temperature that I practice is *not* in the yeast's comfort zone.

OG matters because it tells us about the expected alcohol content of the final product. Roughly speaking, the percentage of alcohol can be found by subtracting 1,000 from the OG and then dividing by 10. More precisely:

$$\text{abv}(\%) \approx 0.105(\text{OG} - 1000). \tag{2.1}$$

This empirical formula works for me and will work for you if your mashing efficiency is similar to mine. If you don't have a hydrometer to check OG, you can still estimate the alcohol content from the weight of grains used in the mash. Assuming that you are making a 6-gallon batch of beer and that you measure the weight, W, of the grain in pounds, then the percentage of alcohol will be approximately

$$\text{abv}(\%) \approx 0.53W, \quad W \text{ in pounds.} \tag{2.2}$$

You will obtain a higher alcohol content if you include a lot of sugar, since sugar is 100% fermentable, whereas grain is, at most, 62% fermentable. However you make the estimate, treat it as only a rough guide; there are so many variables in the brewing process that you can brew the same brew twice in the same way and get different OGs and different alcohol contents.

Before pitching the yeast into the wort, it is a good idea to aerate the wort first. So take a sterilized jug and place it under the faucet at the bottom of your fermentation vessel. Open the faucet to let out the wort. Let it drop some distance, and splash into the jug. Then pour the jug contents back into the fermentation vessel[13]—from a height, if you can. The aim here is to churn up the wort as much as possible, to oxygenate it. The boiling process will have removed air from the wort, but the yeast needs oxygen to reproduce efficiently; hence the aeration. We aerate the wort only at the last minute before pitching, however, because bugs also like oxygen.

Now you have 6 gallons of oxygenated sugary water at a comfortable temperature that is just perfect for growing germs. We won't let that happen. As soon as you have finished aeration, open two sachets of

13. Remember to shut off the faucet first—but you knew that.

brewer's yeast and sprinkle the contents on the surface of the wort.[14] Place the fermentation bin lid on top but don't fasten it down. Now wait for 10 minutes; you can pass the time by marveling at the action of the yeast. You have just poured about six billion of the little beasts into your wort. Many of them will not be viable, but those that are alive and kicking will, even as you contemplate them, be waking up and checking out their new environment. I will get into the dynamics of yeast populations during the next chapter; here, we can simply admire these microorganisms. Once they have checked out your wort, they spend a few hours manufacturing the right kind of enzymes so that they will be able to feed off the sugars that they find in the wort. This first, inactive stage of the fermentation process is often called the *lag phase* by brewers, since nothing is happening outwardly. After 10 minutes, you stir the yeast into the wort (with a sterilized spoon). Place the lid on top, loosely as before, and let your phalanxes of yeast cells do their thing. Apart from maintaining the wort at around 68°F (20°C) there is nothing else that you can do to help them; they are on their own for the next several days.

I always get anxious at this stage because the brewing process is out of my control. Is the yeast viable? Will bugs establish themselves first, turning the wort into 6 gallons of saliva? Is the temperature right? I should know better by now, since the yeast almost always knows what it is doing. If, by some mischance, there is no sign of fermentation after 12 hours, then pitch in another couple of yeast sachets. For our current brew there are no such problems, and within a few hours of pitching you peek under the lid[15] and observe some changes to the surface of the wort. The top-fermenting yeast you have used have started to gather at the surface, as a brownish scum. This scum contains other gunk that the yeast has brought up with it, but much of what you see is yeast. The

14. This method, of sprinkling yeast directly onto the wort, works perfectly well. It is probably a good idea—and many homebrew books and Web articles recommend this—to rehydrate the wort beforehand. Rehydration consists of nothing more than sprinkling the yeast onto some warm (body-temperature) water half an hour before pitching.

15. Don't breathe while you peek. Contamination is a constant threat at all times after the boiling phase. So avoid breathing your buggy breath over the wort and keep the lid on (but loosely).

yeast cells are reproducing fast and using up the supply of air and sugar that you carefully provided. After a further few hours (perhaps 6–24 hours after pitching—times vary greatly with yeast, wort, and brewing environment) the scum thickens into brown foam that will persist for half a day or so. The foam is generated by carbon dioxide gas, CO_2, which the yeast generates as a fermentation by-product. (We will look at the details of yeast action in a later chapter.) Then the foam changes: fissures open up the brown scum, and creamy white foam breaks through. For a while, the surface of the wort is multicolored, brown and white—particularly if you are brewing a dark beer—but the white dominates eventually and lasts for a day or two. It gradually subsides, having generated copious amounts of gas (which is why you leave the lid loose, instead of fastening it tight, so that the CO_2 can vent; otherwise the lid would blow off and create an almighty mess).

You can get some idea of these three stages of primary fermentation—scum formation, billowing foam, foam collapse—from figure 2.6, taken from one of my brews. Note in particular the ring of scum left behind by the retreating foam. This can produce off-flavors in the beer if left too long, so all the homebrew gurus recommend skimming this scum; this is a practice that I urge you to follow. Using a sterilized spoon you dutifully remove the gunk from the surface rim of the wort—now turning into beer, since the primary fermentation process has begun to generate alcohol. You place the lid back on, loosely, and withdraw for a few days to let the yeast continue with its work. The dramatic opening movement of this symphony is now over, and the rest of the piece is conducted at a sedate pace—an adagio. Fermentation continues, and CO_2 is still being produced, but the pace has slackened.

SECONDARY FERMENTATION

A few days after pitching the yeast (between 2 and 6 days, in my experience) it is time to transfer the beer to a carboy for secondary fermentation. You should probably use your (sterilized) hydrometer again to see that the gravity of the beer has been reduced sufficiently. The standard rule of thumb is to transfer the beer when the gravity has decreased to 1020 or below. By this stage, most of the fermentable material has been turned into alcohol and water, plus a lot of gas (we will calculate just

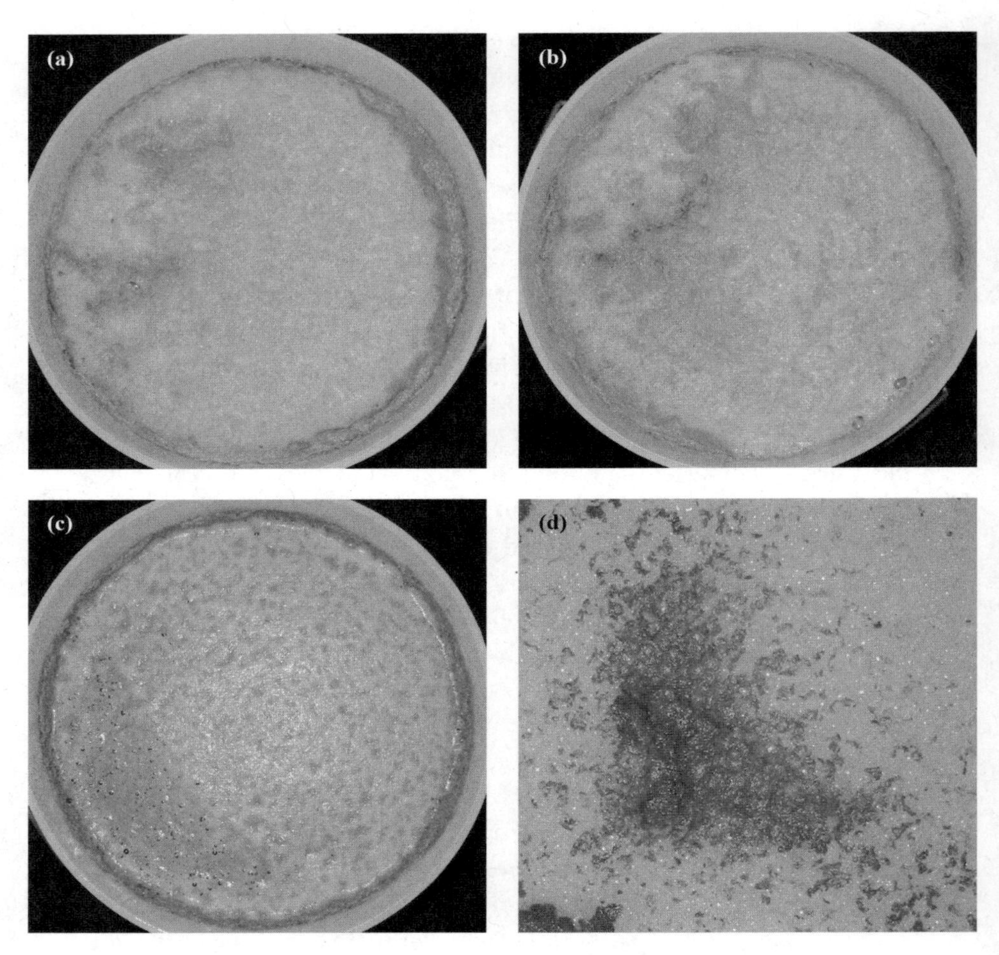

Figure 2.6. Primary fermentation: (a) brown foam, giving way to (b) creamy foam, after which (c) the foam collapses, leaving a ring of scum that should be skimmed; (d) a more detailed view of the brown and white foam.

how many bubbles are produced in chapter 5). The yeast is feeling the pinch; raw materials are becoming scarce and the fermentation process slows. In practice I rarely bother with the hydrometer and just transfer the beer once the fermentation has subsided. With a little experience, you will know when the time is right without needing to take a hydrometer reading. Anyway, the timing is not crucial. Let us say that you decide, after 4 days, to transfer the beer to a carboy for secondary fermentation. How do you transfer the beer, and why?

First, the why. We need to keep the beer away from air. While we deliberately introduced air just before pitching the yeast, from here on we need to exclude air because it encourages contaminating bugs. Air can better be excluded when the beer is in a carboy than when it is in the fermentation bin, with its loosely fitting lid. Another reason is to get rid of the trub, or gunk, that has accumulated at the bottom of the fermentation bin—take a look; you will see a layer of sediment. Generally speaking, brewers (and wine makers too, for that matter) don't like their beverage sitting on trub for very long because of the risk of introducing off-flavors. So you use a plastic hose to siphon off the beer into a carboy, and stop siphoning when the beer is gone and only the trub is left in the fermentation bin. Now you attach an air lock (sometimes known as a *fermentation lock*) to seal the carboy. This lock keeps air out but allows CO_2 to vent. In figure 2.7 you can see a plastic carboy with an air lock fitted. I note, incidentally, that a few homebrewers object to plastic carboys: they prefer glass. The claim is either that plastic imparts an unwanted flavor to the beer, or that it is porous, allowing air to seep through. In my experience, neither of these claims holds water—so use plastic; it is less expensive and more robust.

Second, the how. The standard method, which I practiced successfully for many years, is to siphon the beer from bucket to carboy with a plastic tube, as just mentioned. This method minimizes the amount of air introduced into your beer—recall that, from now on, we are supposed to exclude air to reduce the risk of contamination—and is convenient enough, though you must be careful about cleaning the siphon tubing both before and after using it. Now that I have told you what you are supposed to do, I will tell you what I actually do, though no doubt some purists will haul me over the coals for suggesting it. I open the faucet of the fermentation bucket and let the beer drop into the carboy. This is quicker and involves no messing around with siphons. It also undoubtedly introduces air. I thought about this for a while, some years ago, and decided that aerating the beer at this stage might not be such a bad idea. After all, fermentation is to continue, and so long as there is fermentable material in the beer, our yeast will consume the

Figure 2.7. Purists would object that I have allowed too much air into this carboy; they would have topped it up with water to the neck of the carboy so as to exclude air. In this case, because the beer had plenty of fermentation left in it, the gap would soon be filled with CO_2, thus driving out the air and reducing (indeed, practically eliminating) the risk of contamination. The air lock tells us when the fermentation ceases.

oxygen that we have added. So I risked a batch of beer and tried the dropping method. It worked fine—the beer did not go off, even after several months—and I have used this system ever since, with no problems whatsoever.

One advantage of a glass carboy is that it permits you to see clearly how the fermentation is progressing. Bubbles that accumulate at the top surface tell you that the secondary (slower) fermentation is under way; when these bubbles disappear, then fermentation has essentially ceased. It stops when the raw materials (sugar and oxygen) that are consumed by the yeast have been used up. During secondary fermentation these products are in short supply, and the yeast cells have converted to a different mode of operation—outlined in chapter 4—than they practiced when times were easy. Over the next week or two you watch the airlock on the top of the plastic carboy. At first this will bubble quite frequently—perhaps several times a minute, depending upon the gravity of the beer when you transferred it to the carboy. As the days go by, the bubbling rate will slow, and slow, and slow. After a week or two, check the gravity with a hydrometer. If it is below 1010, secondary fermentation is deemed to be completed,[16] and you move on to bottling your beer.

BOTTLING

Three weeks have elapsed since you first pitched the yeast—or perhaps half that time, depending upon temperature, wort content, and the friskiness of your yeast. The signs of fermentation in the carboy have all but disappeared, and you decide that it is time to transfer your beer into bottles. Here we come to a choice that is largely a matter of personal preference. Some homebrewers prefer glass bottles, and some prefer plastic. Some prefer screw-on caps, and others corks or those clamp-on stoppers as seen, for example, on bottles of the Dutch *Grolsch* beer. Some prefer none of the above, and opt for cans. I prefer one-liter plastic bottles with screw caps; this is the most economical option, and the most convenient if managed with due care and attention.

First, we need to clean the bottles. This is a tedious business (I use about 10 gallons of water for every gallon of beer brewed, because so much cleaning is required: bottles, carboys, siphons, fermentation buckets, etc.), and so I will delegate it to you. (Well, you need to learn, and what better way than by practice?) There are standard cleaning and sterilizing supplies to be obtained from any homebrew store. You obtain some chlorinated detergent powder and make up a solution; you rinse all your bottles with this solution; you rinse again, twice, with clean water. If you are feeling virtuous, you may apply a solution of potassium metabisulfite (a sterilizing agent) as well, and rinse and rinse. Confession time: I rarely use both; life is too short. Despite this laxity, I have never lost a bottle of beer through contamination. So, you decide how

16. Of course, this point is rather arbitrary; bottling when the gravity is a few degrees higher or lower than 1010 is quite acceptable. Secondary fermentation continues for several weeks, and our cut-off point serves merely to provide some measure of assurance that there is not much fermentable material left in the beer.

virtuous you want to be. One helpful hint about cleaning plastic bottles: resist the temptation to use a bottle washer. These implements score the inside of the bottle, creating crevices in which bugs would lurk. Far better to rinse a bottle as soon as you have poured out the beer, to prevent sticky deposits that are difficult to clean. If you are not able to rinse immediately—perhaps The Guys are at your place enjoying a few of your brews and would scoff mercilessly at your domesticity—then put the cap back on to avoid evaporation, and rinse later.

However you do it, let's consider it done and move on. You have 23 clean plastic one-liter bottles and a carboy full of green beer, which shows no sign of fermentation. The yeast has run out of food, and the little fellows are feeling miserable.[17] First you add a glassful of water to a jug, and add priming sugar. How much sugar you add depends upon how fizzy you want your beer, but you'll use somewhere between 6 and 12 ounces of sugar for a 6-gallon batch of beer. Because I hale from the English brewing tradition, and because I do not filter my beer, I tend towards the lower end; after much experimentation I have settled upon 7 ounces of sugar. (Why should the amount of priming sugar depend upon whether or not the beer is filtered? All will be revealed soon enough.) Some experts recommend boiling the sugar solution for a few minutes first, and I generally follow this practice, though occasionally I don't boil it and have not noticed any difference in the final product. The second solution we need to prepare (again, you will do all the work for me) is a fining agent. A half teaspoon of gelatin in a glassful of warm water, stirred until dissolved, if you please. Now add this to the sugar solution in the jug. Again, I sometimes omit the gelatin. It does seem to help clarify the beer, which is a little turbid with suspended yeast and other particles. On the other hand, when I omit the fining, the beer clears anyway—it just takes a week or two longer to do so. You decide to play safe, and so you make up the gelatin solution and add it to the jug. Now you divide the jug contents among all 23 bottles as equally as you can.[18]

17. Well, I imagine that the yeast cells are feeling a little low. They are starving, their friends are dying all around them, and they are wallowing in their own excrement—enough to ruin anyone's day. Yeast excrement, by the way, is alcohol, an observation that may or may not cause you to pause for thought.

18. With my last few brews I have experimented with a simpler method of adding

I find the next stage very rewarding, but I suppose that you should be allowed to do it, since you have done all the hard work. Siphoning the beer into the bottles gives a great sense of satisfaction, of a job completed, of fruitful labor that will soon reap benefits. You must siphon, this time, rather than simply open the faucet because the bottle top is narrow, you really don't want to add yet more air at this stage, and your carboy does not have a faucet. There is a certain skill that develops when bottling beer: at first you are clumsy, and the beer froths up in the bottle as you siphon, but you quickly learn to tip the bottle and pour slowly. (Raising the bottle so that it is nearly at the level of the beer in the carboy will slow the flow considerably.) When all 23 (22 if you spilled some) bottles are filled, screw on the caps, finger tight.

Store the bottles of beer upright in a cool room where the temperature does not vary too much. Brown plastic bottles keep out harmful light, but my bottles are green and so I play safe and store them in a dark cupboard.[19] Over the next few weeks, the yeast in your beer (yes, despite your separating the beer from the trub, there will still be a few billion yeast organisms suspended in the beer) will eat up the priming sugar, generating a little more alcohol and, more importantly, carbonating the beer (*conditioning the beer*, as the brewers say). It is important not to overdo the priming sugar, because the bottle might explode. If you like your beer very frothy—12 ounces of priming sugar rather than my 7 ounces—you may encounter a problem with unfiltered beer. Because we condition our beer in the bottle, it naturally generates some sediment. In time, this sediment settles on the bottom, and when we pour out the beer, we must be careful not to disturb it. When poured successfully, by a skilled expert with years of dedicated experience, the beer in the glass is crystal clear, and you have an inch or so of turbid liquid left in the bottle. This inch is lost beer—a libation to the gods,

---

priming and fining solution (anything to simplify the brewing process): add the solution directly to the carboy instead of to the bottles. This is certainly simpler, but results are mixed. It can stir up sediment in the carboy—the last thing I want to do—and this sediment makes the beer more turbid so that it takes longer to clear. Also, the solution is not uniformly distributed throughout the beer; some bottles get more and others less. On balance, I will probably revert to the method outlined in the main text.

19. You must await chapter 6 to learn why bottle color matters.

Figure 2.8. A cold pale ale. Despite being cooled in a fridge before serving—very refreshing on a warm summer's day—it has plenty of flavor because it was full-mash brewed using a lot of pale malted barley and only a little sugar.

thanking them for providing us with a beautiful tipple. If poured too quickly, the beer will foam up in the glass and will also transfer some sediment, which clouds the beer and makes it look unappetizing. The problem with highly carbonated beer is this: when you uncap it, the rising bubbles can stir up the trub even before you begin to pour. So, no matter how carefully you pour from bottle to glass, your beer will be turbid. So, if you *really* want very fizzy beer, you may be obliged to filter it and carbonate artificially, as the MBs do when manufacturing Macroswill. I will not tell you how to do this because I have never done it and because I would be leading you down the road to perdition.

The amount of beer froth and bubbles that 7 ounces of sugar can generate is quite sufficient (see figure 2.8 for a freshly poured glass of my pale ale) and yet does not cause sediment to stir when the bottle is first uncapped. Incidentally, the caps on plastic beer bottles are very good these days and hold the carbonation well for months. I recently consumed a stout left over from the previous winter—over a year old—and it was still fully carbonated. Some mild beer that had been in the bottle for 8 or 10 months was similarly in full pomp when poured. You do not have to wait this long, of course: your lovingly brewed beer will be ready to drink after three weeks in the bottle. (Less time, and the full flavor will not have developed; more, and the beer will mature—become drier, less fresh, and a little less hoppy, but very good even after months. Note from figure 2.1 that I include *time* as one of the key ingredients of the bottling stage; it is hard to overdo the amount of time that you put into your beer.) One advantage of plastic bottles over glass is that you can feel whether the beer is in condition—that is, whether it has carbonated: the internal pressure of gas noticeably stiffens the bottles.

Three

Yeast Population Dynamics

Give me a woman who loves beer and I will conquer the world.

—Kaiser Wilhelm II (1859–1941)

And the bartender says to Rene Descartes, "Another beer?"
And Descartes says, "I think not," and disappears.

—Alfred Bester (1913–1987)

MULTIPLICATION BY DIVISION

Now that you know how to make beer at home, I can begin to show you how physics determines the way in which beer is constructed. In other words, I can describe to you how certain aspects of the brewing process can be understood mathematically. Given the number of parameters that influence beer production and the consequent variability of brewing, you will appreciate that any mathematical description will necessarily be approximate. My aim here is not to write a treatise but to show you how physics can be applied to beer and beer making. For those of you who like mathematical details, a technical appendix at the end of this chapter summarizes the math development in this chapter in enough detail for you to reconstruct all the steps of my analysis. Those of you who would rather chew a box of thumbtacks than read math can omit the appendix and still understand the basic ideas propounded here if you are prepared to take my math on trust. (All the equations are explained in words.)

I have said something about the nature of brewer's yeast in the last chapter and will say some more in the next chapter. Here we concentrate upon the population of yeast cells in the wort, from the moment they are pitched until the moment that the beer is bottled. The manner in which this yeast population grows and then declines can be understood from simple physical principles, and from a few such principles we will construct a mathematical model of yeast population that roughly accords with what is known from experiments and what we can observe from brewing beer at home.

PITCH IT RIGHT

Yeast is yeast is yeast, right? You know by now that this is not the case. Though closely related, the many strains of *Saccharomyces cerevisiae* (see fig. 3.1) have differing characteristics. More distant cousins such as bread yeast can be used to make beer, as we saw in chapter 1, but it would not be beer as we know it, captain. Other distant cousins cause disease in humans. Siblings such as top-fermenting ale yeast and bottom-fermenting lager yeast have different characteristics and produce very different beers, as we have seen. Even within the top-fermenting yeasts that we employed to make our beer in chapter 2, there are many different strains that produce different beer characteristics. I have said it several times before: yeast contributes half the flavor of beer. Different strains ferment to different degrees the many different types of sugar that are present in wort and produce small quantities of by-products that flavor the beer. These by-products are manufactured during the fermentation process in different amounts. Furthermore, the various yeast strains behave differently in wort of a given temperature and react differently to changes in temperature.

Given the variability of brewer's yeast, it is easy to understand why commercial brewers sometimes guard their proprietary yeast jealously. For example, Guinness used to make a bottle-conditioned beer, containing live yeast, from which a homebrewer could extract the yeast and make a stout that approximates Guinness. No longer. The point is that by propagating a strain of yeast for many generations, brewers can select cells (say, yeast strain X) that yield a desired flavor or character of beer, and then consistently produce this beer by pitching only yeast X into their wort. So, each commercial beer might have its own strain of yeast. It is not difficult to change the character of yeast by such

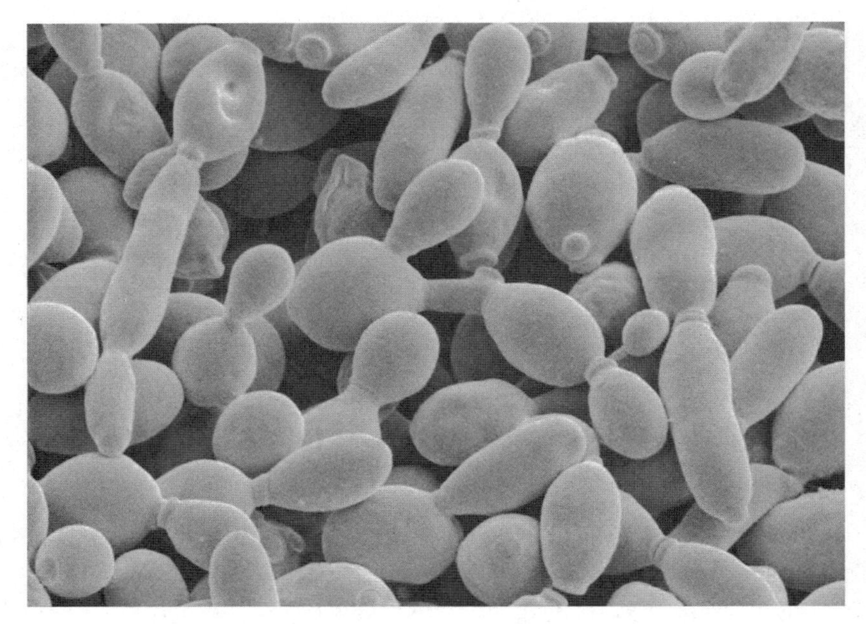

Figure 3.1. Saccharomyces cerevisiae, or brewer's yeast, magnification 4000×. I thank David Scharf for providing this image.

"selective breeding." Look how dogs and cattle have evolved under human guidance, and these creatures can breed only once a year or so. Yeast, under ideal conditions, can double in population every 2 hours, and so it doesn't take long to generate new strains. The difficult trick, I suppose, is to maintain the same strain for decades.

The yeast that most homebrewers pitch into their wort is supplied dry, in small 5- to 7-g sachets (see fig. 3.2). The quality and viability of such dry yeast has improved immeasurably from the bad old days in the 1970s when homebrewing first took off.[1] Nowadays we can depend upon the yeast being genuine brewer's yeast, and being viable. A 7-g sachet will contain a mere four billion cells, however, and this is not nearly enough to start a brew. The experts consider that yeast den-

1. I recall being given a pint of homebrew at a party, during my student days in Edinburgh, Scotland, in the mid-70s. The beer was flat and turbid and had a greenish hue, with a quarter-inch of sediment at the bottom of the glass. It smelled of sour apples and tasted like a mixture of apple juice, toothpaste, and vinegar. It isn't often that I pour away a nearly full pint of beer.

sity when first pitched into the wort should be about ten million cells per milliliter of wort (i.e., 10^7 mL^{-1}; there are about 16 mL/in^3), whereas our 7-g sachet pitched into 6 gallons of wort corresponds to only 200,000 mL^{-1}. Even my doubled-up amount corresponds to a *pitching rate* of only 400,000 mL^{-1}, which is about 25 times short of the optimum.

What does it matter, you may ask. After all, if the yeast cells can double in number every 2 hours, then the factor of 25 can be made up in less than 10 hours—five generations. Well, it isn't crucial (or we would always pitch in 175g of dry yeast, rather than 7g), but a shorter interval of rapidly reproducing yeast is probably better. To understand why, we need to appreciate a little of how yeast does its stuff. We will learn in the next chapter that brewer's yeast cells are *facultative anaerobes*. (I'll bet you've been saying the same thing for years.) A facultative anaerobe can respire with or without oxygen. In an oxygen-rich environment such as the initial wort that you have assiduously aerated, the yeast cells practice aerobic respiration—which, as we will see, is an efficient method of breaking down the wort sugar to supply the energy that the yeast cells use for reproduction. When the oxygen supply dwindles, the cells switch to anaerobic respiration, which is less efficient but works. Not all yeast types can do the anaerobic trick, but every strain of brewer's yeast can do it—has to do it—because only anaerobic respiration produces alcohol. So, the brewer would like his yeast cells to get to the anaero-

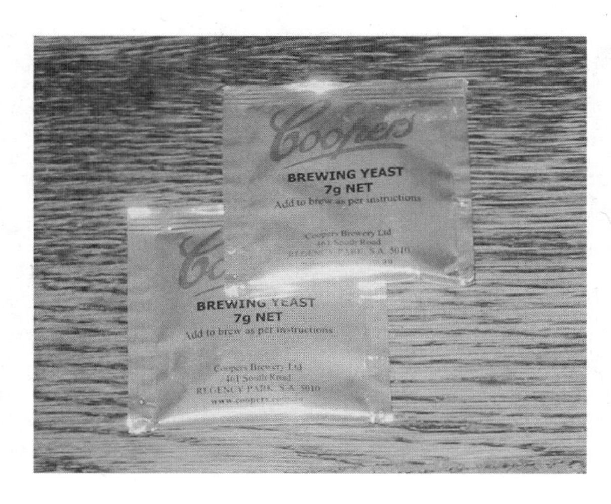

Figure 3.2. Sachets of dry brewer's yeast. At least two sachets should be used for pitching into 6 gallons of wort.

bic stage as quickly as possible. Another reason that we would like to get over the aerobic phase quickly is that aerobic respiration leads to yeast cells' generating fermentation by-products that influence the flavor of the beer, even in small amounts. This is not necessarily a bad thing, but in large amounts these by-products are undesirable: the beer tastes wrong. The compounds that give rise to such off-flavors are esters (which smell or taste of banana), fusel alcohols (solvent), diacetyl (butter, butterscotch), and acetaldehyde (green apples).[2]

To avoid the risk of off-flavors, we like to limit the number of generations that the yeast cells spend in our beer. The experts tell us that three to four generations is the best number. They say that the maximum density of yeast cells in the wort for an efficient fermentation is about a hundred million per milliliter (10^8 mL^{-1}). Recall that the recommended pitching rate was ten million per milliliter, and that the reason this density was chosen is that it takes three or four generations for the yeast to get up to a hundred million per milliliter. Once there, they find the wort is getting crowded and oxygen running low, so they switch to anaerobic respiration and start generating alcohol. Now suppose we start with my recommended minimum of two 7-g sachets of yeast. The yeast will take four or five generations—say 9 hours—to reach a density of 10^7 mL^{-1} and then another three or four generations to reach the maximum 10^8 mL^{-1}. That is, instead of three to four generations of aerobic respiration, our yeast will spend eight to nine generations at it, in the wort. The amount of by-product will be greater, and so the risk of generating off-flavors will be higher. (For an accessible account of brewer's yeast replication and aging, see Powell et al.)

Another question arises in your fertile mind: if aerobic respiration is bad for the beer, then why do I encourage you to aerate the wort? Surely if the yeast were pitched into unaerated wort, the little buggers would get down to anaerobic respiration more quickly, resulting in more alcohol and fewer off-flavors. The problem is that for the yeast to convert all or most of the wort sugars into alcohol (to achieve a *high attenuation* of

2. The German top-fermented wheat beer *Hefeweizen* tastes fruity, not because fruit is added to the beer, but because of esters that are intentionally a part of the beer. My pale ale, when young, has a fresh aroma of apples—presumably due to a small amount of acetaldehyde.

the wort), there needs to be the maximum density of cells. So unless you pitch a vast amount of yeast into your wort, the pitched yeast cells will need to reproduce. (Otherwise, not all the sugars would be converted to alcohol, and you would be left with a *stuck fermentation*.) But with little oxygen in the wort, the yeast can't reproduce much, and there would be insufficient yeast to convert all the sugar to alcohol. The resulting beer would be weak and sweet. How much yeast would you need to pitch if you wanted to start at the maximum density of 10^8 cells per milliliter? About 4 pounds of dry yeast—say 250 seven-gram sachets, which is economically and practically unfeasible.

In the previous chapter I recommended that you use yeast that was produced during the fermentation of one batch of beer to start off the next batch of beer—the brewing equivalent of chain-smoking, I suppose. Now you can see why: not only are these yeast cells old hands at converting your type of wort, but there are a hundred million of them per milliliter. (OK, fewer of them are alive and well, but far more than you would obtain from a couple of sachets.) In fact, experts recommend that you use the yeast from the top of your wort for this purpose, rather than use the stuff that has dropped to the bottom.[3] They say that you should wait a day or so after pitching, so that the cell density is at a maximum, and then save some wort from the top for the next brew. This recommendation sounds good to me: compared with using the leftover yeast at the bottom of a batch, you get a higher percentage of viable yeast and less trub (dead cells and other products that have dropped out of the wort). In practice both methods work well, and much better than pitching with a couple of dry yeast sachets. I make sure that if I use the leftover sludge at the bottom of the fermentation bucket, I do so within a day (less risk of contamination or of the dead yeast cells' decomposing).

3. One μb we consulted (the Kona Brewing Company on the Big Island, Hawaii) told us that they use yeast from batch n to start the fermentation for batch $n + 1$, for $n = 1$ to 9. After that, they throw away the yeast and start with fresh yeast. This is because yeast that has undergone several fermentations begins to mutate and also picks up bacterial contaminants, and so is only good for a few batches. For my homebrewing, I conservatively stick to $n \leq 3$ because I lack the efficient and sophisticated cleaning equipment of a professional μb.

Figure 3.3. German *Hefeweizen* wheat beer, filtered (*left*) and unfiltered (*right*). There is little doubt that the filtered beer looks more attractive, but beerophiles frown upon the process.

With all these solids in the wort, you may well wonder how it is possible to make beer that is bright, i.e., crystal clear. Commercial brewers usually filter their beer before bottling it or putting it in barrels because the product looks more attractive to customers than the unfiltered beer (see fig. 3.3). I don't filter my homebrew, and most craft brewers do not filter their beers, because filtering diminishes the flavor. You may recall that we remove the trub by simply pouring the beer off the sediment twice (after primary and secondary fermentation), and then perhaps we permit ourselves the use of natural fining agents to encourage the process. There remain in the beer a few zillion dead or dormant yeast cells, plus other suspended solids. Sometimes these cloud the beer, and sometimes not (usually not, especially after the beer has matured for a few weeks), depending upon the type of beer and the brewing technique. If brewed properly, most beer will exhibit little or no cloudiness. The possibility of a little haziness—in the beer, not the brewer—is a small price we pay for better beer.[4]

4. Wheat beer is always cloudy unless filtered (see fig. 3.3). My pale ale develops a slight chill haze if put in a refrigerator, but not otherwise. My bitter beers are crystal clear; the stout and mild beers are so dark you can't tell if there is a haze or not.

MATHEMATICAL MODEL OF YEAST POPULATION DYNAMICS

There are two variables in my simple model of yeast population evolution: the number of active yeast cells and the supply of nutrients. By "active" cells I mean the population p that is alive and kicking: these cells use up nutrients—they are not dead or dormant. By "nutrients" I mean food and oxygen, taken together, here denoted by F. The ratio of these numbers, $D = F/p$, is the density of nutrients—the supply of food and oxygen that is available to each yeast cell.

The mathematical model is constructed from two quite simple physical assumptions. I will outline these assumptions here and then plot the model prediction for the evolution of yeast cell population as the fermentation process develops. We will see that the predictions are roughly in accordance with the observed evolution of wort during fermentation. Our first assumption concerns the rate at which wort nutrient is used up. The nutrient supply decreases with time: the longer the time interval, the more food gets eaten. Also, the nutrient supply decreases faster for a larger yeast population. Both these assertions seem eminently reasonable and can be combined as the first assumption of our model. Mathematically:

$$dF = -\gamma p \, dt \tag{3.1}$$

The constant of proportionality is γ. In words:

Over a short time interval the nutrient supply falls, by an amount that increases proportionally with the time interval and with the yeast population.

Now for the second assumption, about the rate at which yeast cell population changes. The cell population changes with time and changes faster for a bigger population. The cell population increases with yeast fecundity and decreases as food density falls. These assumptions can be combined into the following mathematical statement:

$$dp = \left(\beta - \frac{\alpha}{D}\right)pdt \qquad (3.2)$$

In words:

> *Over a short time interval the yeast population changes by an amount that increases proportionally with the time interval and with the yeast population. The increase in population is greater for fast-reproducing strains, and the decrease is greater when the food density is low.*

In equation (3.2) the Greek letter β represents yeast cell fecundity or reproduction rate, which we can fix by observing the time it takes for a population of cells to double. There are three Greek letters in my equations (3.1) and (3.2)—α (alpha), β (beta), and γ (gamma)—and all three of these parameters are assumed to be constant: they do not change with time.[5] This is undoubtedly a simplification of the real-world situation. For example, it is likely that the rate at which yeast cells reproduce depends upon available nutrients, which depend upon time. If we accept the simplification, however, then we obtain a simple mathematical model that we can work with, rather than a more complicated model that is unsolvable. This is acceptable so long as we remember that simplifications have been made and so the model predictions can only be regarded as approximate.

From equations (3.1) and (3.2) we can predict how the wort nutrient density $D(t)$ and yeast population $p(t)$ change with time. The math is worked out in the appendix to this chapter and leads us to the following predictions:

$$D(t) = D_0 \exp(-\beta t) \qquad (3.3)$$

5. This is what a scientist means by the word *parameter*. It is a constant factor that influences the outcome of a system but is not a *variable* of the system. So, for example, in a system that consists of a swinging pendulum, the time is a variable (because pendulum motion changes with time, and time varies), whereas the pendulum length is a parameter (because it is constant for a given pendulum, but changing it leads to different pendulum motion). For our yeast population model, time is the variable and α, β, and γ are parameters.

$$p(t) = p_0 \exp(\beta t - r(\exp(\beta t) - 1)), \qquad r = \frac{\alpha}{\beta D_0} \tag{3.4}$$

Here D_0 and p_0 represent the initial nutrient density and the initial cell population at time $t = 0$, when the yeast is first pitched into the wort. In words:

> *Nutrient density falls exponentially as time goes by; the cell population increases exponentially at first, but then crashes catastrophically when the available nutrients become too meager to support the population.*

The population evolution prediction, equation (3.4), is plotted in figure 3.4 for two different pitching rates.

If we fed nutrients into the wort at just the right rate to keep the level constant all the time (at a value F_0, say), then the evolution of yeast cell population would be very different. It would show the same initial exponential explosion, as in figure 3.4, but then it would level off to a constant value (of $p = \beta F_0 / \alpha$, as can readily be shown from our second assumption, eq. [3.2]). In our case, however, the initial nutrient levels are not replenished. The nutrients get used up and eventually reduce to zero. (This fact is utilized in the appendix to show that the constants α and γ must be equal.) The food disappears, and the yeast population disappears soon afterwards, as the population starves. In fact, not all the yeast cells die—most of them go into a dormant, inactive state, hoping for improved conditions—but the point is that the active population decreases markedly. In figure 3.4 you can see this effect: a disastrous drop in population after a sharp peak. The drop in population is probably not represented very accurately by my model because it takes no account of how the cells respond to the reduced nutrient levels: in practice they recognize that resources are getting short and so they switch from aerobic to anaerobic respiration, as we have seen. Nevertheless, we can take the graphs of figure 3.4 as indicative of the level of yeast *activity* as time goes by, from the moment that the yeast cells are first pitched into the wort.

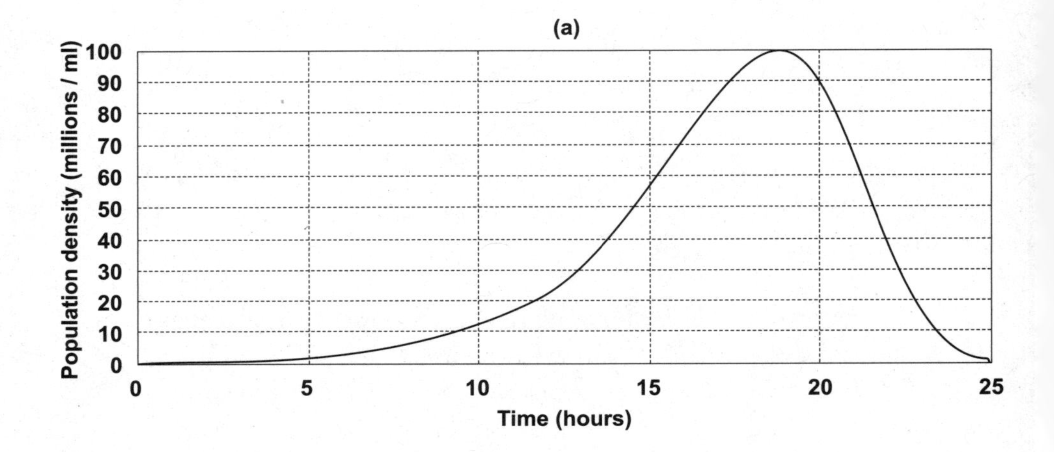

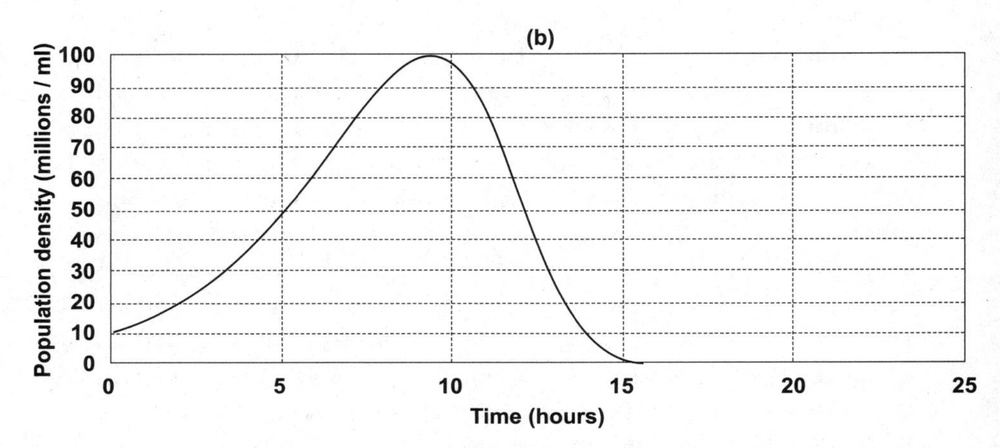

Figure 3.4. A simple mathematical model of yeast population density (millions of cells per milliliter of wort) vs. time (hours) at two pitching rates: (a) pitching rate (initial population density) of 400,000 mL^{-1}; (b) pitching rate of 10,000,000 mL^{-1}. These plots may be inaccurate for the anaerobic phase of fermentation (after the population peak) because the model does not distinguish between aerobic and anaerobic yeast cell respiration.

ASIDE: GOOD AND BAD PHYSICS

Given that, under ideal conditions, yeast cells reproduce every couple of hours, we can estimate from the figure how many generations must pass before the cells switch to anaerobic respiration and start to make alcohol. If we pitch two sachets (a pitching rate of 400,000 mL^{-1}, recall), eight or nine generations go by before resources start to get scarce; if we pitch at the recommended rate (10^7 mL^{-1}), only three or four generations are required. This prediction matches what we are told by the experts. So we can attach some credence to our simple model. Our mathematical modeling of yeast population evolution is typical of how physicists progress. They make a simple theoretical model of a process that they want to understand, fit some of the model parameters to the observed data, and then see what predictions the model makes. Here we took from the observed data of yeast action two facts: the reproductive rate of yeast cells under ideal conditions and the peak population in the wort. We then applied the model to predict *when* the cell population reaches a maximum. (We see in fig. 3.4 that the time required depends upon pitching rate.) Checking against observation, we see that these predictions are about right. For example, my homebrew wort does nothing for a few hours after pitching two yeast sachets, and then it generates copious foam (aerobic respiration), which subsides within a day. If I pitch yeast from an earlier brew, the process is accelerated: virtually no lag phase, then rapid foam generation, which subsides within about 12 hours.

Were we serious about accurately modeling yeast population dynamics, we would now refine our model to include the yeast responses to reduced nutrient levels. Our model would evolve, becoming more sophisticated and more accurate in its predictions. So long as the model is capable of making predictions, we can consider it to be of value: it can be tested against experimental observation, the very cornerstone of science. A bad model has lots of different parameters, each of which needs to be determined by observation, and makes no predictions. For example, suppose that our yeast population model had 24 independent parameters ($\alpha \ldots \omega$) instead of two (α, β). To determine these 24 parameters we would need 24 observations: on the rate at which yeast cells reproduce, the maximum population density, the time of the popula-

Figure 3.5. A reward for studious effort.

tion peak, etc. So what would be left for the model to predict? Everything would be fixed by the parameters, and so this 24-parameter model would have no predictive power. The model has too much "wiggle room," in that any new data that come along just require adjusting some of the many parameters in order to fit the data. With our two-parameter model we required only two observations, and the third fact (the time of peak population) emerged from the model unambiguously, which shows that the model describes what is actually going on; we are not just fitting data.[6]

I have belabored this point, but it is an important one that is sometimes forgotten, especially by inexperienced scientists: a valid model of

6. This philosophy goes by the name of "Occam's razor" (a.k.a. "the principle of parsimony"). To decide which of two theories is the better: if they both describe observations equally well, then choose the one with fewer parameters.

a physical process must have predictive power in order to describe the underlying dynamics. Otherwise it merely describes data.

This chapter has been quite mathematical, especially for those of you who choose to work your way through the appendix. As a reward for your efforts, you are now welcome to treat yourself to one of my brews (see fig. 3.5).

APPENDIX: MATHEMATICAL DETAILS OF THE YEAST POPULATION MODEL

From the definition of nutrient density ($D = F/p$) we see that a small change in density is given by

$$dD = \frac{dF}{p} - \frac{F\,dp}{p^2}\,. \tag{3.5}$$

Substituting for dF and dp from equations (3.1) and (3.2) yields a differential equation for D:

$$\frac{dD}{dt} = \alpha - \gamma - \beta D. \tag{3.6}$$

We expect that, once the food supply is exhausted ($D = 0$) there will be no further change in D; it will stick as zero. For this reason, we see from (3.6) that $\gamma = \alpha$. Integration then yields

$$D(t) = D_0 \exp(-\beta t), \tag{3.7}$$

which is the exponential behavior described in the main text. Now substitute equation (3.7) into (3.2) to obtain a differential equation for p, and integrate with the initial condition $p(0) = p_0$ to obtain equation (3.4). Incidentally, this equation looks very much like the *extreme value distribution* known to statisticians.

We can fix the two independent model parameters α and β (or, equivalently, r and β) as follows. We see from (3.4) that the yeast cells initially grow exponentially in numbers: this is *Malthusian growth*. Observation of real brewer's yeast under such conditions shows that they double in number every 2 hours, which means that the fecundity parameter must be given by $\beta = ½\ln(2)$, i.e., $\beta = 0.35\ \text{hr}^{-1}$. The second model parameter, r, is obtained by noting that the peak yeast cell density is observed to be about 10^8 mL^{-1}. So the ratio of pitching rate to peak density is about 0.004 for the pitching rate of figure 3.4a (and is 0.1 for the pitching rate of figure 3.4b). It is easy to show from equation (3.4) that this ratio is given algebraically by *er*, where e=2.7182818285 is the base of natural logarithms, and so $r \approx 0.00147$ ($r \approx 0.037$). Hence the two parameters of the model are fixed.

It is worth noting here that the time at which cell population peaks can be calculated from equation (3.4). It is $t_{peak} = \ln(r)/\beta$. For times exceeding t_{peak}, or perhaps a little earlier, real yeast cells switch to anaerobic respiration and so the value of β will change. A more detailed model of yeast cell population dynamics would include this effect.

Four

Brewing Thermodynamics

Beer is proof that God loves us and wants us to be happy.
—*Attributed to Benjamin Franklin (1706–1790)*

I work until beer o'clock.
—*Stephen King*

THE FERMENTATION REACTION

Catabolism is the name given to the multistep process by which cells (here yeast) break down complex compounds (sugars) to form molecules with less stored energy (carbon dioxide, water, and alcohol). Energy that is released by breaking up the sugars is moved to the cells' powerhouse—the mitochondria—as the chemical ATP. Now I am not a biochemist, and my duty here is to tell you about the technology and physics of brewing, so I will summarize the complicated biochemistry of fermentation with just a few words.

Sugar comes in many forms. The simplest is glucose ($C_6H_{12}O_6$), consisting of 6 carbon, 12 hydrogen, and 6 oxygen atoms. Other sugars consist of two or more glucose molecules strung together. Thus maltose, the most important sugar that is released from grain during the mashing process (recall chapter 2), consists of two glucose molecules and so is a *disaccharide* sugar. Other sugars (*polysaccharides*) consist of more than two glucose molecules bound together. Many sugars can be converted by yeast into glucose; this requires the yeast cells to produce enzymes which perform the necessary chemical transformation.

The yeast can then act upon the glucose to produce the energy that is needed for reproduction.

Not all sugars are fermentable, which is to say that the yeast cannot produce the specific enzymes that are required to convert some sugars (lactose, a disaccharide, is one example) into glucose. The lag phase of fermentation—the first few hours after you pitch the yeast into the wort, you may recall from chapter 3—consists of the yeast cells' getting to know their new chemical environment. One of the tasks they perform during this phase is to manufacture the enzymes that they will need to convert the sugars that they find in the wort.

After the yeast cells obtain a Ph.D. in biochemistry they break down glucose molecules to scavenge the energy. The basic reaction (though this is, in fact, a multistage process) is as follows:

$$C_6H_{12}O_6 + 6O_2 \rightarrow 6CO_2 + 6H_2O + 688\ \text{kcal mol}^{-1}.$$

In words:

> *Glucose plus oxygen produces carbon dioxide gas plus water plus a lot of energy.*

Note the presence of oxygen: this is *aerobic respiration*, which is the yeasts' preferred way of breathing. The clever little fellows can breathe without the presence of oxygen, as we will see, but aerobics works best for them. Now, to give you a feeling for how much energy is released by the sugar during aerobic respiration, consider the following. *If* all the energy were released as heat, instead of being bound up in ATP molecules to power yeast reproduction, one pound of sugar would raise the temperature of 6 gallons of water by 139°F (77°C)—enough to kill the yeast. In this scenario, the 6-gallon batch of homebrew we brewed in chapter 2 would become steam rather than beer!

This degree of heating does not occur; the sugar energy goes elsewhere. Most of the energy that is obtained by the yeast breaking apart sugar molecules is utilized to make more yeast, and so they move into the growth phase, whereby cells multiply exponentially, as we saw in chapter 3. At some point they use up all the oxygen that we have provided for them by aerating the wort, and the yeast cells then switch to *anaerobic respiration.*

The basic anaerobic reaction is

$$C_6H_{12}O_6 \rightarrow 2C_2H_5OH + 2CO_2 + 28 \text{ kcal mol}^{-1}.$$

In words:

Glucose is incompletely broken down to produce ethanol (alcohol) plus carbon dioxide gas plus a little energy.

This is the fermentation reaction. If all the energy released by anaerobic respiration of one pound of sugar were available for heating up 6 gallons of wort, the wort temperature would increase by about 5°F (3°C).

Of course, yeast cells produce energy from sugar for their own use, not ours. They are not perfectly efficient machines, and so some small fraction of the sugar energy is in fact released as heat. In the sections that follow, we will calculate how the temperature of a 6-gallon batch of homebrew evolves in time due to this "yeast heating." First, though, I would like to show you how thermodynamic considerations can determine some time scales of more immediate relevance to the pragmatic brewer. (Once again, you can read around the math if you so choose, but there is a beer waiting for you at the end of the chapter if you stick with it.)

HEATING THE WATER

Now we are in a position to apply some simple thermodynamics to the brewing process. I will concentrate upon homebrewing because that is the brewing that I know most about and because I have described the process in chapter 2. The thermodynamics of brewing on an industrial scale are rather different; I will discuss these differences at the end of this chapter.

How long does it take to raise a certain volume of water—4 gallons is the typical volume utilized during mashing—to the temperature range appropriate for mashing, say 150°F (65°C)? We can calculate the time taken if we know the power of our heating element, a few other parameters, and some basic thermodynamics.

The plastic bucket that I use for mashing is provided with a 2-kW heating element. Let this power be represented by P. The amount of heat that it delivers to the water during a small time interval dt is

$$P\,dt = mc[dT + \alpha(T - T_a)dt]. \tag{4.1}$$

Here m is water mass, c is the water's specific heat (the amount of heat needed to raise the temperature of one gram of water by 1°C), dT is the increase in water temperature during dt, T is water temperature, T_a is ambient (room) temperature, and α is the "cooling coefficient" for my plastic bucket—representing how fast heat is lost through the sides of the bucket. I have measured this parameter to be $\alpha = 0.20\ \mathrm{hr}^{-1}$. In words, equation (4.1) says:

The element heats the water by a small amount during a short time interval, but some heat is lost through the sides of the bucket.

The hotter the water, the more wastage. This makes sense; it is in accordance with our everyday experience. Equation (4.1) can be written as a differential equation

$$\frac{dT}{dt} = \frac{P}{mc} - \alpha(T - T_a), \tag{4.2}$$

which any mathematician can solve easily[1] to obtain

$$T(t) = T_a + \frac{P}{\alpha mc}\,[1 - \exp(-\alpha t)]. \tag{4.3}$$

This solution tells us how the water heats up with time and is plotted in figure 4.1 ("Phase 1"). It is important to note that equation (4.3) is the

1. Technically, this differential equation is separable and is easily integrated. The initial condition assumed is that the water temperature equals ambient temperature, T_a, at time $t = 0$.

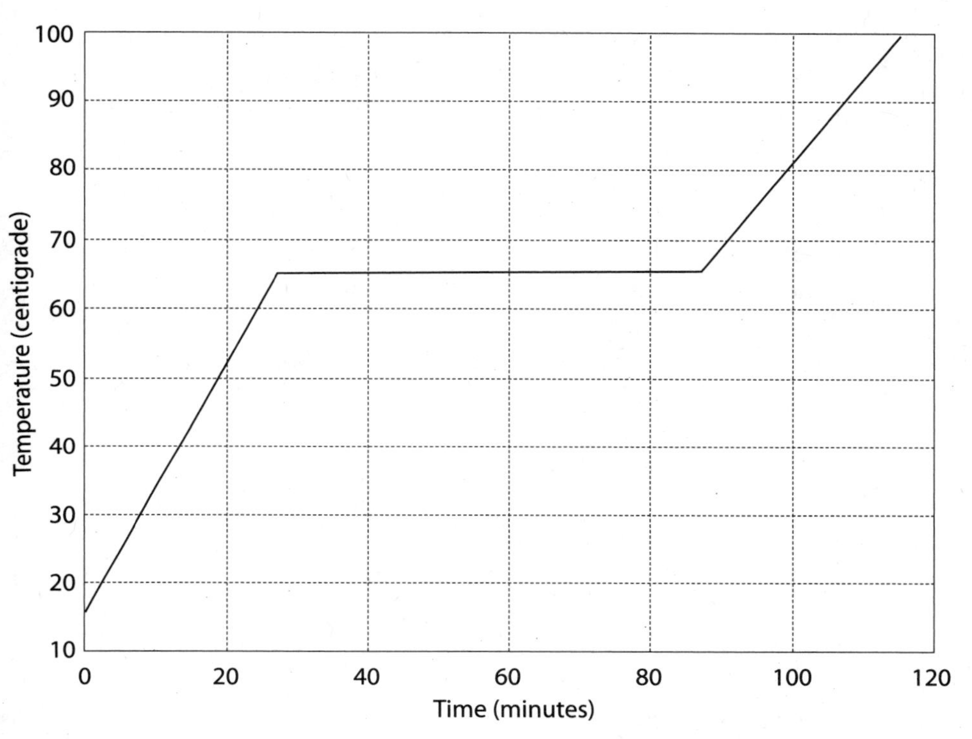

Figure 4.1. Wort temperature (°C) vs. time (minutes). *Phase 1:* The wort is heated and temperature rises from ambient (here assumed to be 15°C) to mashing temperature (65°C). *Phase 2:* The temperature is held constant during the 1-hour mash. *Phase 3:* The wort is again heated until it boils. Heating is less rapid during this last phase because the wort volume is increased from 4 gallons to 6 gallons at the end of phase 2.

same as equation (4.1); we have used no additional physics, only math, in order to get from (4.1) to (4.3). All the physics that we needed to arrive at the answer is contained in the assumptions underlying (4.1).

If mash temperature is $T_m = 150°\text{F}$ (65°C) then we can determine the length of time it takes our element to heat the water to T_m, from equation (4.3):

$$t_m = \frac{1}{\alpha} \ln\left(\frac{P}{P - \alpha m c (T_m - T_a)} \right). \tag{4.4}$$

Putting in the numbers we find that the time taken for my 2-kW element to heat up 4 gallons of room-temperature water to 150°F is about 27 minutes. Lo and behold, observations show that this is indeed about right.

So now I don't have to stand around the mash bucket and measure the water temperature every few minutes. I can go away and prepare the grain, knowing that I have a little under 30 minutes to do so.

Once the mash is completed we have to boil the wort (now made up to 6 gallons). How long does this take? A calculation very similar to that I have just described leads to the following expression for the interval t_{mb} to raise the wort from mash temperature T_m to boiling point T_b:

$$t_{mb} = \frac{1}{\alpha} \ln \left(\frac{P}{P - \alpha mc(T_b - T_m)} \right). \tag{4.5}$$

For the same parameters as before (except that we are now heating 6 gallons instead of 4) we find that $t_{mb} = 29$ minutes (see fig. 4.1, "Phase 3"). So, I need not stand over the bucket waiting for water to boil, a proverbially tedious business, but can instead get on with other tasks for half an hour or so; there is usually a lot of cleaning and tidying up to do during the process of beer making, as my wife is wont to remind me.

So why is this interval important? After all, the wort is to boil for an hour, so why do I need to know when it starts to boil? Sure, I need to know t_{mb} and add an hour, to know when to turn off the heat and permit the wort to cool, but there is a more immediate reason. I need to attend to the wort a few minutes before it begins to boil because of another aspect of beer physics. The wort begins to foam and bubble as the temperature approaches boiling point, and this foam becomes very sticky and persistent. The sugar content may account for this stickiness. The bubbles rise to the surface, come into contact with cooler air, and become glutinous (imagine bubbles of toffee). The wort becomes covered with a head that is sometimes so thick and stiff that it is almost like meringue; further heating causes this head to rise and boil over the side of the bucket, creating a sticky mess that is a real pain to clean up. So to prevent overflow, I keep an eye on the wort for a few minutes before boiling starts and, if necessary, stir the forming head into the wort.

In fact, I try to avoid stirring if possible, not for any pragmatic reasons to do with beer production, but because the evolving wort surface is a fascinating sight for a physicist. As the temperature approaches the boiling point, a brown scum forms on the top of the wort, which becomes sticky as described, forms a solid cap, and rises slowly as the pressure builds up beneath. It never fails to bring to my mind the thought of earthquakes and volcanoes. The molten magma beneath a thin, solid crust pushes against the crust due to internal pressure, causing it to rise, slowly at first, but more rapidly prior to an eruption. The "wortquake" erupts as fissures that split open, revealing creamy white foam beneath (unless I am obliged to break up the foam by stirring). Sometimes instead of fissures, a large bubble arises and bursts, issuing forth a small eruption of "volcanic" steam. Once fissures split the surface, the surface foam is swallowed quickly by the boiling wort, and the danger of overflow is past. (Photos of these three phases of wort boiling are shown in fig. 4.2.) Now I can get on with the cleaning for an hour.

COOLING TO PITCH TEMPERATURE

Many homebrewers like to cool the wort quickly following the boil. There are two reasons for this. First, as mentioned earlier the wort is vulnerable to bacterial contamination once it cools below about 160°F (71°C), and so you really want to get it down to pitching temperature (68°F–99°F, or 20°C–37°C) as quickly as possible. This is the reason usually quoted in brewing instructions. Recall from chapter 3 that bacteria can multiply up to six times faster than can yeast cells, so that if bacteria get into the yummy, nutritious wort before our yeast, they will take it over and we will end up with 6 gallons of vinegar, or worse.[2] Thus, conventional wisdom states that we should cool our wort quickly and pitch a lot of yeast into it when it has cooled sufficiently, so that the yeast will take over the wort and muscle out the few airborne bacteria that inevitably will find their way in. The second reason is the so-called *cold break:* a sudden drop in wort temperature will cause some suspended proteins to sediment out. Because these proteins may account for some off-flavors if left in, it is probably a good idea to get rid of them.

2. If you are talking to somebody while staring into your bucket of cooled wort—perhaps you are arguing with your wife about the house smelling like a brewery—and some of your spittle gets into the wort, then you may end up with 6 gallons of spit. If your child sneezes into the wort then . . . yuck.

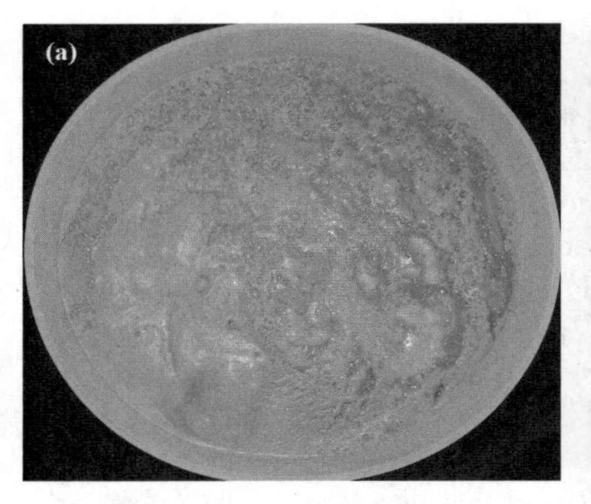

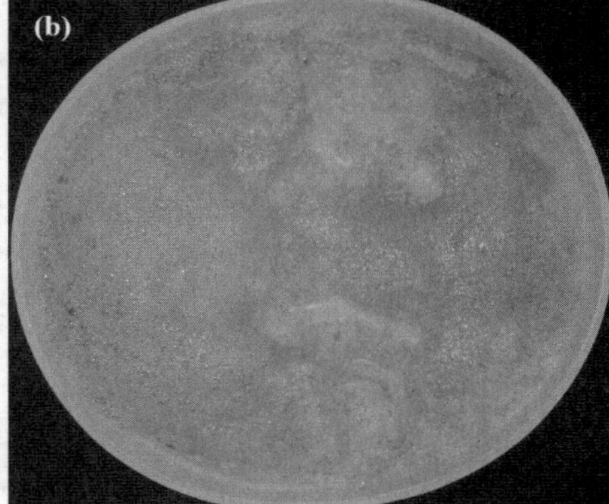

Figure 4.2. The three phases of wort boiling: (a) formation of scum and sticky froth; (b) foam breakthrough; (c) rolling boil.

I find that placing a lid on the bucket of cooling wort as soon as the boiling phase is completed allows the wort to cool naturally (i.e., slowly) without risking contamination. There are no living bugs in the wort when I put the lid on because the wort is at boiling point, and no airborne bugs can get through the lid from outside. So forget about the first reason. Now what about the cold break? I find that no off-flavors develop if I let the wort cool naturally. I have brewed about 80 batches of beer over the years, and not one has developed off-flavors. So, forget about both reasons and let your wort cool naturally.[3] How long will that take?

3. You may have the equipment (say a spare refrigerator) to enable you to accelerate the cooling. If so, by all means use it and let me know what difference it makes. You may also accelerate cooling by placing a towel soaked in cold water around the

We can again use equation (4.2), except that now there is no heating element. So set $P = 0$ to obtain

$$\frac{dP}{dt} = -\alpha(T - T_a), \tag{4.6}$$

and solve this differential equation[4] to yield figure 4.3 and

$$t_p = \frac{1}{\alpha} \ln\left(\frac{T_b - T_a}{T_p - T_a}\right). \tag{4.7}$$

Here t_p is the time required for the wort to cool from boiling point to pitching temperature, T_b is the boiling temperature of water (100°C), T_p is the temperature at which we pitch our yeast, and T_a is, as before, ambient temperature (say 59°F, or 15°C). We find that t_p depends upon pitching temperature, as shown in table 4.1. Thus, if we are prepared to pitch our yeast into 95°F (35°C) wort, we must wait 7.2 hours for the wort to cool, whereas if we use a pitching temperature of 68°F (20°C), we must wait a further 7 hours. Personally, I go for the higher pitching temperature, as discussed in chapter 2. The brewing literature disagrees, generally affirming that my yeast will die, or at least not reproduce well, at 35°C, but the brand of dried yeast that I use seems perfectly happy to get on with the job at such temperatures. Anyway, the wort cools further during the lag phase.

By extending the physics of equation (4.1) a little, we can calculate how much of our wort will evaporate during the boil:

$$P\,dt = \varepsilon m c_v\,dt + mc[dT + \alpha(T_b - T_a)dt]. \tag{4.8}$$

---

bucket. Such evaporative cooling can be very effective because each gram of water that evaporates takes away 540 calories of heat. Another favored method of cooling is to plunge a length of coiled copper tubing into the wort and run cold water through the tube. I am not so sure that this is a good idea. By bringing the tubing into physical contact with the wort, you are, ironically perhaps, increasing the risk of contamination. But many people use this method successfully—your choice.

4. This time the initial condition is temperature $T_b = 100$°C at $t = 0$.

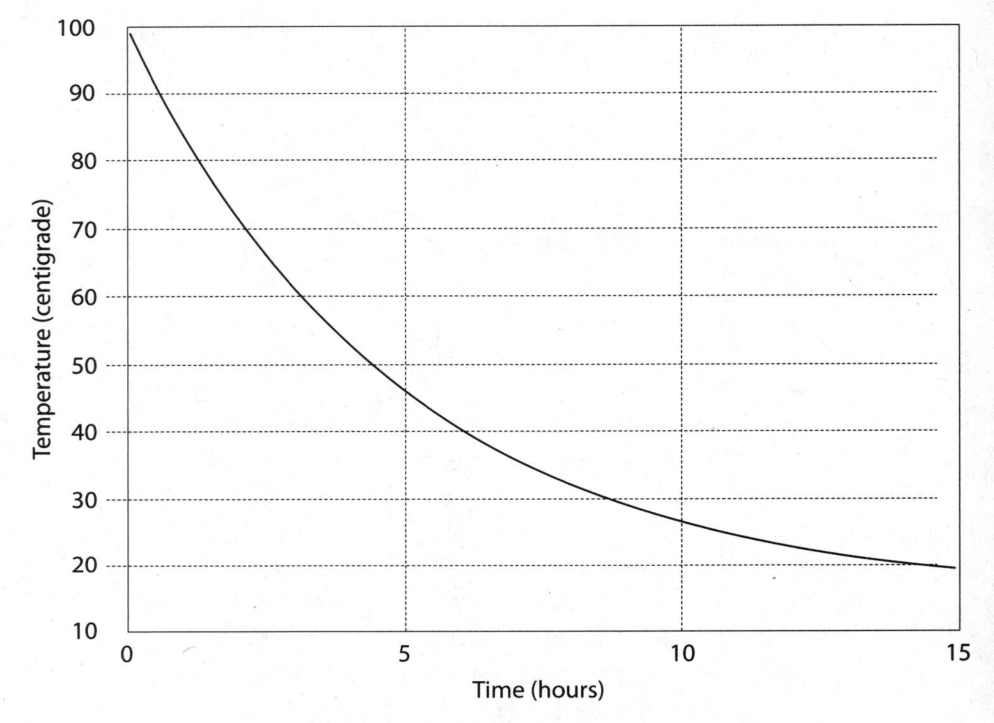

Figure 4.3. Wort temperature (°C) vs. time (hours): cooling phase.

We have added a term to the right-hand side of the equation: c_v is the latent heat of vaporization for water, 540 cal gm^{-1}. (Note that I am assuming wort to have the same thermal properties as water.) The term ε is the fraction of water mass that is evaporated. In words:

> *The heat supplied by our heating element goes into evaporating water plus raising the water temperature, as well as heat lost through the bucket wall.*

During the boil phase, the change in temperature is zero. Thus we find

$$\varepsilon = \frac{P - \alpha mc(T_b - T_a)}{mc_v} , \qquad (4.9)$$

Table 4.1 Time for wort to cool from 100°C to various pitching temperatures

Pitching temperature, T_p (°C)	Cooling time, t_p (hours)
35	7.2
30	8.7
25	10.7
20	14.2

which works out at about 11% per hour. Again, from my observations of the actual wort boiling phase this number seems to be about right: I lose a little more than half a gallon of wort for a one-hour boil.

HEATING DUE TO YEAST

So far, my discussion of beer thermodynamics has been restricted to the early phases of brewing: mashing, boiling, and cooling of the wort. The next phase is primary fermentation when, recall from chapters 2 and 3, yeast is pitched into the cooled, aerated wort and allowed to replicate exponentially. When the wort becomes crowded with yeast cells and the oxygen and sugar become scarce, the yeast cells switch to anaerobic respiration, producing alcohol. I hinted earlier that yeast respiration causes heating of the wort; though most of the energy released by catabolizing sugar is utilized to make more yeast cells, a small fraction is released as heat. How does the temperature of the wort change as a result of yeast activity? I can once again apply basic thermodynamics to this problem and so obtain a rough estimate, an educated guess, as to what is going on.

As a first step, permit me to make the simplifying assumption that yeast population density is constant. That is to say, the number of yeast cells per cubic centimeter of wort does not change. We know from chapter 3 that this is not the case, that the yeast population explodes initially, but I will make the assumption anyway because it will prove to be instructive. It might be a plausible assumption in the context of industrial-scale brewing once the yeast population has reached an optimum density because in this case the brewers might add just enough oxygen and other resources to maintain such a density, at least for a while, to hasten the fermentation process. However, homebrewers cannot monitor and control their beer production to anything like the same extent. We just pitch the yeast and let it do its own thing: light the blue touch-paper and back off. Later I will relax the unrealistic assumption and derive a "temperature profile" of the wort, plotting the changes in temperature as the yeast population rises and falls.

Let us denote the density of yeast cells within the wort by ρ, the Greek letter *rho* (our *r*), so that my simplifying assumption becomes $\rho = p/V$ = constant. Here p is yeast population and V is wort volume. It is reasonable to assume that heat is generated throughout the volume of the wort but is lost only through the surface, S, the top of the wort and the sides of the fermentation vessel. Mathematically we can express this assumption as follows:

$$dQ = mc\,dT = \gamma\rho V\,dt - \delta S(T - T_a)dt. \tag{4.10}$$

As before, dT is a small change in temperature, and dt is a small time interval. In words:

> *The heat released by yeast catabolizing sugar during a short time interval increases wort temperature throughout the volume, but some of this generated heat is lost through the surface.*

The amount of heat lost increases as the wort temperature rises above ambient temperature, T_a. In equation (4.10) I have introduced more Greek letters: γ (Greek *gamma*, our *g*) here represents the yeasts' ability to generate heat throughout the wort volume, and δ (*delta*, our *d*) stands for the insulating properties of the fermentation vessel.[5] Thus

5. Why Greek? I suppose because physicists and mathematicians run out of English (actually Roman) letters and because the ancient Greeks started us off, mathematically. In this book, I often adopt Greek letters for unknown variables or, as here, for unknown parameters. (You will have gathered that the α parameter here is different from the α of chapter 3. I suppose that I might have chosen a different Greek letter, but there is little risk of confusion. And I like α—it is a cool letter.) More generally, physicists have developed conventions that certain letters stand for certain things. For example, ρ or σ are often used to denote density, and c denotes the speed of light

small δ means that little heat is lost through the vessel walls, and large δ means that a lot of heat leaks out. In fact the constant $S\delta/mc$ is just the cooling coefficient α of equation (4.1).

The rest is just math. As before, all the physical assumptions have gone into our initial equation. From (4.10) we obtain a differential equation and solve it. I will skip this stage—interested readers can fill in the blanks themselves. The result is that the wort starts off at pitching temperature and ends up at a temperature

$$T_y = T_a + \frac{\gamma\rho V}{S\delta} \,. \tag{4.11}$$

I have included the subscript y to indicate that this change in temperature is due to the action of yeast, distributed uniformly through the wort. We don't know what values to assign the parameters γ, δ and so cannot say exactly what the final temperature will be. However, from equation (4.11) we see that T_y increases with yeast density and wort volume. Also, T_y increases more for well-insulated fermentation vessels with small surface areas.

This dependence upon geometry—in particular, that temperature is proportional to V/S—pops up in other areas of biophysics. Thus, warm-blooded animals that live in cold climates are generally larger than warm-blooded animals in warm climates. For example, polar bears are the biggest type of bear. Geese that breed in the Arctic are larger than geese of the same species that breed further south. The reason for this phenomenon, which is well-known to biologists, is contained in equation (4.11). To stay warm, it helps to have a large volume (since body heat is generated throughout the body) and a small surface area (since heat is lost through the surface). Another example: sea mammals tend to be large and well insulated (small δ). This is because water conducts heat better than air, and so without good insulation and a lot of generated heat, whales and walruses and sea lions would soon chill. For the same reason these creatures tend to be more spherical than land mam-

---

or, as here, specific heat. The doubling-up of letter use rarely causes confusion; we know from context what a letter represents. Thus, when Einstein tells us $E = mc^2$ we know he is not talking about specific heat.

mals, since a sphere is the geometrical shape with largest ratio of volume to surface area, V/S.

Yeasts are not walruses, however, as you may have noticed (and I certainly would not want my wort filled with a trillion walruses), and now it is time to relax the somewhat unrealistic assumption about constant yeast density. Certainly, homebrewers cannot easily control the yeast population, so now I will assume that the yeast cells replicate and die away as we determined in chapter 3 (equation 3.4). The evolution of wort temperature for a batch of homebrew, after pitching the yeast, is governed by the following differential equation:

$$\frac{dT}{dt} = \frac{\gamma}{mc}\, p(t) - \alpha(T - T_a) + \frac{P_{hb}}{mc}\,. \tag{4.12}$$

Here $p(t)$ is the time-dependent yeast population that we calculated in chapter 3, and P_{hb} is the external power that I apply to the fermentation bucket during the first few hours after pitching the yeast. In words:

> *The wort temperature increases due to yeast activity and due to external heating; all the while heat is being lost to the cooler environment.*

The external source is a heating belt wrapped around the lower end of the bucket that delivers about 30 W of power. Equation (4.12) is difficult to solve analytically, and so I will turn to my computer to number-crunch a solution, which is plotted in figure 4.4. The important point to note is that wort temperature fluctuates as a result of yeast action and that the initial application of external heat helps to keep the temperature within the yeast comfort zone. Once fermentation is well under way—after, say, 12 hours—I switch off the heat and let the yeast cells do their thing; they need no further help from me.

Some homebrewers adopt a more hands-on approach; they want to control the wort temperature so that their yeast is perfectly happy at all times. If your aim is to produce competition-quality brew, then you may need to get more involved than I do, with my minimalist approach. My beer is good, sometimes very good (if you will forgive the bragging), but probably not gold-medal stuff. If a gold medal, rather

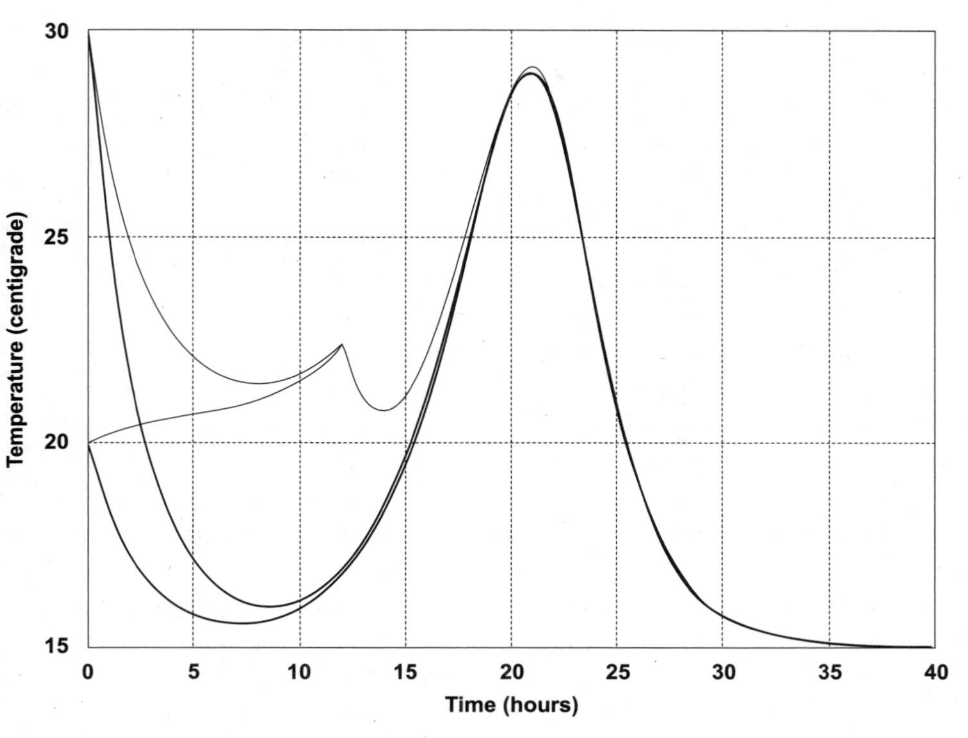

Figure 4.4. Wort temperature (°C) vs. time (hours): primary fermentation phase. Here the pitching temperatures are 20°C and 30°C. Bold lines show the temperature evolution assuming that no external heating is applied. Thin lines show what happens if a 30-W heating belt is applied for the first 12 hours; note how the temperature stays within the yeast comfort zone (roughly 20°C–30°C)—at least until the yeast population crashes.

than amber nectar, is your aim, you may wish to adopt the approach of a number of homebrew enthusiasts and place your fermentation bucket in a bathtub. Fill the tub, and maintain the tub water temperature at the level you want for your pampered yeast. This works because the volume of water in the tub will be much greater than the volume of wort, and so the tub water temperature wins out over yeast heating effects. For such water bath brewing you really want a fermentation bucket that transmits heat readily (thin plastic or metal), and you should stir the water from time to time.

TEMPERATURE REGULATION

One simplification that remains in my thermodynamic calculations concerns the reaction of the yeast to changing temperature. I have assumed that so long as the yeast cells are at a comfortable temperature, they perform at a constant rate. In practice, the rate at which yeast convert sugar depends upon the temperature. More than this: the balance of final products from yeast respiration depends upon wort temperature. Sure, the main products are always CO_2 (which escapes, because it is a gas) and ethanol, but there are other compounds as well (such as sugars that ferment slowly or not at all), produced in small amounts, as we have seen. Some of these trace compounds produce off-flavors, or add body and "mouthfeel" to the beer. Whether for better or worse, they influence the final product. So, my calculations above should be regarded as approximate, since I make no allowance for the finicky behavior of yeast as temperature changes.

The approximation is good enough for the thermal properties of wort, which is what matters in this chapter, but you should be aware that temperature control is very important for making beer. In my opinion, it is the inability of us homebrewers to control wort temperature that limits our ability to produce beer that is consistent from one batch to the next. Consistency is vital for commercial brewers: the public expects to taste the same product each time it buys, say, a pint of Timothy Taylor Landlord or a glass of Anchor Steam. For this reason commercial brewers go to a lot of effort to ensure the consistency of their beers, and much of this effort is expended in ensuring consistent temperatures from one batch to the next. We homebrewers find such temperature regulation much more difficult to achieve, and so our beers vary to a greater or lesser extent, batch to batch.

Water baths may aid consistency after the yeast has been pitched, but prior to pitching we have less control. We have seen that the mashing temperature must lie between about 144°F and 156°F (62°C and 69°C). If the mash temperature falls outside this narrow range, the enzymes that are released by the yeast start to complain. Too low a temperature and enzyme performance is sluggish; too high and the enzymes may be destroyed. Even for mash temperatures within the acceptable range, there are differences. For example, mashing at the upper end of the range will yield beers with higher alcohol content and more body than beers brewed at the lower end of the range, even when exactly the same grain mixture and yeast are used for the two brews. Commercial brewers can take advantage of this temperature sensitivity to brew beers of a specific character because they can exercise exquisite temperature control of their mash and fermentation. We homebrewers cannot, and so, as noted in chapter 2, I simply aim for the middle of the range when mashing (150°F or 65°C). Even then it is possible that some of the wort is above or below the acceptable temperature range because I am unable to maintain a uniform temperature within the wort. To do so would require constant stirring, and there is more to do in life than stand over a bucket stirring warm wort. If you want a gold medal for your beer, then you may be motivated to stir for an hour, but I settle for equalizing the temperature once or twice during the mash. By "equalize" I mean that I draw a few liters of wort out of the faucet at the bottom of the bucket and pour it in the top. This temporarily equalizes temperature top to bottom, which otherwise stratifies, since warmer liquid rises above cold liquid. Even so, there exist temperature variations *across* the wort, especially when I use a lot of grain in the mash (to make, for example, a stout or other strong beer), since the wort consistency is more like porridge than like water and wort circulation is impeded.

For thinner worts (e.g., for lighter beers) the mash can circulate more easily—as in figure 4.5—and temperature variations within the bucket are less marked. Even with thin worts, though, I aim for the center of the acceptable temperature range during the mash phase; such is the variability of mashing in a 6-gallon bucket with a single heating element at the bottom. Again, perfectionist homebrewers may choose to perform their mashing in a water bath.[6]

Finally it is interesting to note the effects of scale upon brewing thermodynamics. We saw earlier that a large fermentation vessel heats up more than a small one because the ratio V/S (equation [4.11]) in-

6. I mean, of course, that the bucket (not the homebrewer) should be in the water bath. But you knew that.

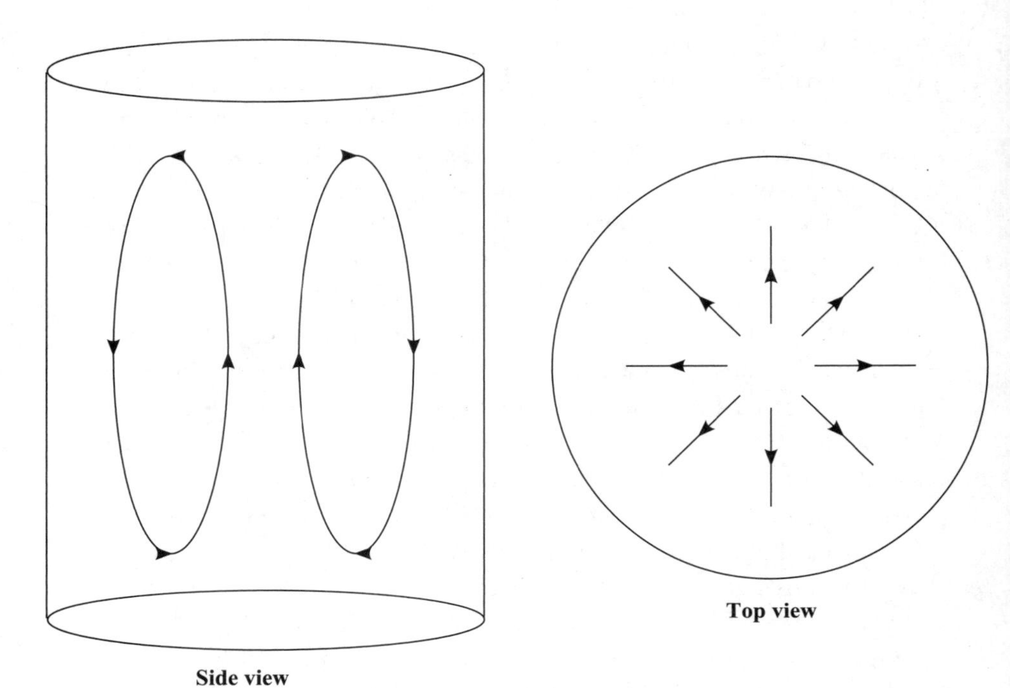

Figure 4.5. Idealized circulation pattern for wort in a fermentation bucket, heated from below. The bucket sides are cooler than the center, and so wort in the center is less dense. The less dense material rises and is replaced by wort descending the bucket sides. Because of this circulation a temperature gradient arises; the wort at the top can be much warmer than the wort at the bottom. If the temperature gradient is not "equalized" (see text), the temperature range can exceed that of the yeast comfort zone (144°F–156°F).

creases with size. This fact means that the heating issues that arise during large-scale commercial brewing are different from those that apply to us homebrewers. A fermentation vat with a capacity of 10,000 gallons has much larger volume/surface ratio than does a 6-gallon fermentation bucket, even if both vessels are the same shape. Beer brewed in a large vat will not face the homebrew problem of initial cooling (as shown in fig. 4.4). Quite the opposite: large vessels of fermenting wort generate too much heat and have to be cooled to keep the yeast happy. Many old-fashioned commercial fermentation vessels are shaped so as to have as large a surface area as possible; modern brewers, in contrast, have recourse to modern refrigeration techniques and need not be so

Figure 4.6. Brewing on different scales.
(a) Macrobrewing in Bavaria, Germany.
(b) Microbrewing in Kona, Hawaii.
(c) Homebrewing in the storeroom, my house. Thanks to the Bavarian Beer Federation for (a) and to the Kona Brewing Company for (b).

influenced by scale or geometry when designing their fermentation vessels. In summary: the difference in scale of MB, μb, and homebrewers (see fig. 4.6) produces different heating problems, which are solved in different ways.

For wading through the math: a beer (see fig. 4.7).

Figure 4.7. For mastering the math in this chapter, you again deserve a cool beer. (On the left, a Hefeweizen; on the right, a porter.)

Five

Bubbles

I am a firm believer in the people. If given the truth, they can be depended upon to meet any national crisis. The great point is to bring them the real facts, and beer.

—*Abraham Lincoln (1809–1865)*

24 hours in a day, 24 beers in a case. Coincidence?

—*Stephen Wright*

FOAMING ALE

Bubbles and beer go together like music and dancing. We have seen that bubbles are an integral part of brewing, and yet it is not essential to include them in the final product. Most wine, for example, is flat—not carbonated. For millennia, though, brewers have chosen to carbonate the beverage that they call beer, and today beer without bubbles is unthinkable. In this chapter we will look at three stages of beer production and consumption during which bubbles feature prominently. We have observed already that copious quantities of carbon dioxide are produced during the fermentation process. This gas emerges from the fermenting wort as bubbles. How many? We will estimate the number of bubbles that a standard 6-gallon batch of homebrew will produce during the fermentation process. Well, it's a philosophically important point, I feel, and one that the nation needs to address. The second occasion when bubbles assume center stage is during the pouring process, when beer emerges from a spigot or tap or bottle or can, and into a beer glass. The liquid froth rises and falls in a manner that depends

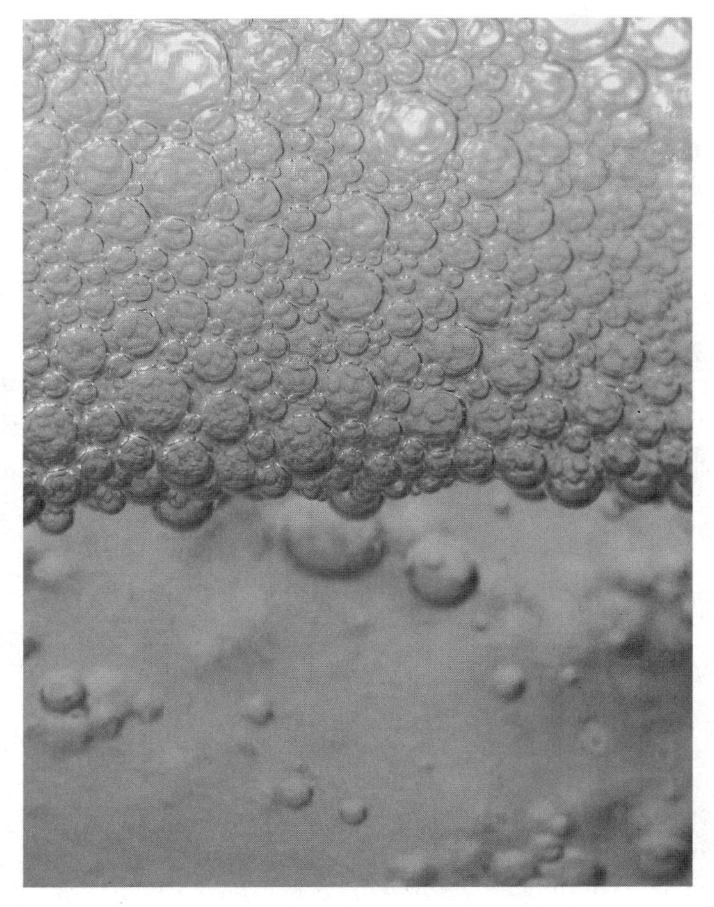

Figure 5.1. Thanks to Martin Eager (http://photos.runic.com) for this image of beer bubbles.

upon the type of beer. It turns out that we can describe the rise and fall of beer foam mathematically; these ruminations will form the subject matter of the middle part of this chapter. We are not alone in such ruminations: the world's MBs have spent much intellectual effort and money on trying to understand beer foam. Finally, beer in the glass produces bubbles that appear on the inside surface of the glass and then rise to the top (and sometimes even *fall*, as we will see). Many a beerophile on many an occasion has stared into his or her pint glass, transfixed by the magic of beer bubbles (see fig. 5.1, to remind yourself what beer bubbles look like). I will enlighten you here, in the final section of this chapter.

FERMENTATION BUBBLES

A typical 6-gallon batch of homebrew (domestic craft beer, if you prefer) will use mostly malted barley to provide the fermentable material, as we have seen. Say we ferment 8½ pounds of pale malt plus one pound of sugar, as in chapter 2. Given that pale malted barley is about 62% fermentable, we see that we have the equivalent of about 6¼ pounds of glucose. To ascertain how many bubbles the yeast cells produce during fermentation, as they chomp their way through this amount of sugar, we need to know how much CO_2 gas they produce. From the aerobic chemical reaction of chapter 4 we saw that each molecule of glucose is converted into six molecules of CO_2. We can calculate the mass of CO_2 produced by comparing molecular weights. A CO_2 molecule has a mass of 44 atomic mass units, whereas a glucose molecule consists of 180 such units. So, each pound of glucose will yield just less than 1½ pounds of CO_2. So 6¼ pounds of glucose gives us about 9 pounds of CO_2.[1]

Now to convert this mass of carbon dioxide into a volume of gas. Air at standard temperature and pressure weighs about 1.2 kg m^{-3} (which is about 1⅕ ounces per cubic foot). So, our 9 pounds of CO_2 occupies about 2.25 m^3 (or 2,250 L—say 80 cubic feet).[2] A typical CO_2 bubble attains a diameter of about 0.5 mm or a little less when it reaches the wort surface (judging by eye, a rough-and-ready method but one that is good enough for our purposes), which means that a typical fermentation bubble has a volume of about 1/15 mm^3. So the 2.25 m^3 of gas is emitted as about 35 billion tiny bubbles.

Given the approximations that I have made for this calculation, I wouldn't trust the exact value calculated, so let us say that a standard 6-gallon batch of homebrew produces between 10^{10} and 10^{11} bubbles

1. I am assuming that most of the gas is produced by aerobic respiration. This approximation will give us a reasonable estimate for the number of bubbles. If you want an *exact* figure, then you will have to count them yourself.

2. I will leave the detailed calculation for you to work out for yourself. You will need to know that the density of air is about 2/3 the density of CO_2, at the same temperature and pressure.

during fermentation. Extrapolating from this figure and using the statistics for world beer production (see the introduction), we can estimate the number of beer bubbles generated yearly on planet Earth during the beer fermentation process: about 10^{20} bubbles (give or take a few trillion). In other words, the world's beer production generates about 100 billion billion bubbles per annum during fermentation. These bubbles represent about 8 or 9 million tons of CO_2 gas. Some of you may be alarmed by this figure, given that CO_2 is a greenhouse gas. To put it into perspective, note that industrial and other human activity generates about 1.8 *billion* tons of CO_2 in North America alone. Beer accounts for 0.5% of this figure, so drink your pint with an easy conscience.

Recall from chapter 2 that some of the CO_2 that we generated when making our homebrew did not escape into the atmosphere. We added priming sugar to the beer bottles so that the beer would "condition"—i.e., carbonate as it matured. Let us see how much gas is generated during this process. Strictly speaking, in this case we are not interested in the number of bubbles, but instead we would like to know how much gas is in our beer bottle and how much pressure it generates. If 8 ounces of priming sugar is needed for a batch of homebrew, then each bottle receives about ⅓ oz (or 10 g/L). We can safely assume that, in the bottle, the yeast must resort to anaerobic respiration, so that (given the anaerobic reaction shown in chapter 4) each molecule of glucose generates two molecules of CO_2. Thus, 10 g of sugar generates about 5g of CO_2 inside each bottle. This gas is held under pressure in the beer, but when the cap is released the beer foams as the CO_2 is released. At normal atmospheric pressure the 5g of CO_2 will occupy about 2.5 L, or 2½ times the volume of beer. As it happens, this is very close to the pressure that commercial brewers favor for their beers.[3]

POURING BUBBLES

The first thing that you notice when beer is poured into a glass is the foam. Before the color settles, before the aroma percolates, you see the

3. Actually, a little less: commercial brewers like to gas up their beers with between 2.55 and 2.65 volumes of CO_2, according to Probrewers.com, a beer industry online resource. The pressure inside a barrel is typically 10–30 pounds per square inch (psi).

title of this book. Froth, to a greater or lesser extent (and the extent depends upon a number of factors, as we will see), bursts forth exuberantly. First impressions are important, and so commercial brewers have spent a lot of time and effort to ensure that their beer froths, foams, bubbles, effervesces, sparkles, or fizzes in the most appealing manner. Charlie Bamforth, the guru I quoted in chapter 1, reckons that humans have investigated the "fizzics" of beer foam for over half a century. Consult the extensive literature and you will find that the generation and dissipation of bubbles in a glass of beer is generally divided into four contiguous and yet interdependent categories. First we see bubble formation. Then the bubbles rise to form a head, a process known as *beading*, or *creaming*. Then we see the bubbles mature: big ones grow bigger while smaller ones shrink, in a process given the catchy name of *disproportionation* (a.k.a. to physicists as *Ostwald ripening*). Bubbles also mature due to *drainage*, as liquid beer falls out of the head, leaving behind dry foam that consists of polygonal-shaped bubbles. In the next four sections I will discuss the physics that underlies these four different phases of fizz.

Bubble Generation

Beer in the can, bottle, or barrel contains a lot of dissolved CO_2 under pressure: we made it that way. When the beer is poured, the pressure is released and the CO_2 wants to come out of solution. The microphysics of bubble formation is quite complex, however. Even though the temperature and pressure of the beer (when in the glass) is too low to support all the gas that it contains in solution, this gas cannot easily emerge from the liquid on its own. It needs help, in the form of nucleation sites. The sites may be impurities in the beer or, very commonly, on the surface of the glass into which we have just poured our beer.[4] If you are one of that group of people who has spent many hours staring into a pint of beer,[5] then you will already be familiar with the phenom-

4. There is an analogy here with supercooled pure water, which may remain a liquid at temperatures well below the freezing point, in the absence of nucleation sites from which ice crystals can grow. Once such sites are provided—say by throwing filings or other impurities into the water—ice crystals will grow quickly.
5. This group is sometimes given the name "males."

enon of nucleation sites, since you will on occasion have observed a stream of bubbles issuing from one small area of the glass surface, with no bubbles originating from anywhere nearby. The small area contains a spur, or a trapped dust particle, or some other defect in the otherwise smooth glass that permits gas to exit the liquid and fly upwards as bubbles. There is even a formula that physicists have derived to tell you the mean bubble radius, *r*, to emerge from a given nucleation site:

$$r = \left(\frac{3R\gamma}{2\rho g} \right)^{1/3}. \tag{5.1}$$

Here *R* is the nucleation site radius, or length scale. The beer surface tension and density are represented by γ and ρ, respectively, while *g* is the acceleration due to gravity at the earth's surface. Except for *R*, all these factors are either constant or change little from one glass of beer to the next. So, roughly speaking, we can say that beer bubble size varies slowly with nucleation site size and is pretty much constant for a given site.

You may have noticed how beer sometimes *fobs* (foams excessively) when poured into a rough-sided container such as a plastic beaker with many surface scratches. Read "many nucleation sites." You can, in fact, reduce the amount of fobbing by filling the beaker with water, pouring out the water, and then pouring in the beer. The water fills in some of the cracks, thus reducing the number of nucleation sites.[6] You can also reduce the amount of foam by pouring more slowly, since mechanical agitation also influences foaming rate, as any beer drinker who has jiggled a glass of beer will tell you. There is considerable beer lore about the correct way to pour beer from bottle to glass, and this lore arose so that we can pour out a glass of beer with just the right amount of froth on the top. Here is a typical description, with explanatory bubble physics in parentheses (see fig. 5.2):

> *Tip a clean, air-dried glass* (few nucleation sites,[7] to prevent fobbing) *to an angle of 45° and pour the beer slowly* (to avoid mechan-

6. Or perhaps dust particles are the nucleation sites, and the water washes away the dust. See Liger-Belair's book for an extended discussion of bubbles.
7. I use plastic bottles for my homebrewed beer, and these scratch easily, thereby

ical agitation, and to allow time for the head to form) *from the bottle. Gradually raise the glass upright as it fills* (to avoid spilling the beer, idiot). *Finish with a flourish by increasing the distance from bottle to glass as you pour the last of the beer* (to increase mechanical agitation and so generate more froth, so that it projects above the glass, and a little of it spills over the edge).

A good head on the beer is deemed desirable for a number of reasons. It is visually appealing and is what beer drinkers expect to see when they are presented with a glass of beer. Hence the interest of big brewers in foam generation and beer head retention.[8] The froth also presents the beerophile with his favorite tipple in two forms: foam and liquid. He will knowingly sip the foam and then the beer, sensing the different hoppiness of the two. He will lift up his glass and peer critically into it, thus suggesting to those around him that he must know a great deal about beer. He will look at the bubble structure of the foam and hope to see uniform, small bubbles (heads formed thus are deemed to be more esthetically pleasing).

The *shape* of the beer glass is also a factor in determining the amount of froth and its appearance as beer is poured. We will see why this is so after considering the other stages of froth generation and evolution.[9]

Beading

We have seen that nucleation sites and mechanical agitation cause bubbles to emerge from the beer once we have released the pressure by opening a beer can or bottle. We now pour the beer into the glass and observe a greater or lesser amount of foam accumulating on the top of

---

providing nucleation sites. To avoid scratching them, I do not use a bottle-brush when cleaning the bottles; instead I rinse the bottles thoroughly and let them dry naturally.

8. Would it be too cynical of me to suggest that, for Lyte beer manufacturers, this is a case of a bad product disguised with good packaging? I think not.

9. The Belgians, in particular, are aware of the importance of glass shape; they seem to have a different type of glass for each of the many types of beer they brew. Germans allow for foam in the size and shape of their beer glasses. The English tend not to allow for foam because their beer is less effervescent, and anyway their wide pint "pots" tend to disperse rather than preserve the foam.

Figure 5.2. How to pour beer into a glass. Photo courtesy of the Bavarian Brewers Federation, Munich, Germany.

the beer. This beading, or creaming, process has been investigated both theoretically and experimentally by brewing scientists and others. In this section I present some of the basic observations about beer head formation and then provide a simple mathematical model that covers much of the observed behavior of beer foam.

Bubbles evolve once they have formed: they either shrink to nothing or else they grow and squeeze against one another. The behavior and distribution of bubbles in foam is the subject of the next couple of sections. Here we need to note only that the head forming on top of our glass of beer initially consists of small, round bubbles suspended in liquid beer. The liquid drains away quite quickly, leaving larger, polygonal bubbles that are more stable. (These two phases of beer froth are illustrated in fig. 5.3.) Foam also evolves: it rises and then falls as the bubbles come and go. Famously, the decay of beer foam has been shown to obey an exponential decay law. A paper written in the *European Journal of Physics* in 2002 discussed this weighty subject (see Leike), and its author was later awarded an Ig Nobel Prize for his work.[10]

10. The Ig Nobel Prize spoofs the Nobel Prize and is awarded annually at Harvard University at about the same time that the Nobel Prizes are awarded in Stockholm. Many Nobel winners adjudge the Ig Nobels, which are awarded for science achievements that "first make people laugh and then make them think." Past prizes have been awarded for a wide range of crucially unimportant topics. Examples from 2006: Ornithology—U.S. research explaining why woodpeckers don't get headaches; Medicine—a U.S./Israeli medical case report entitled "Termination of Intractable Hiccups

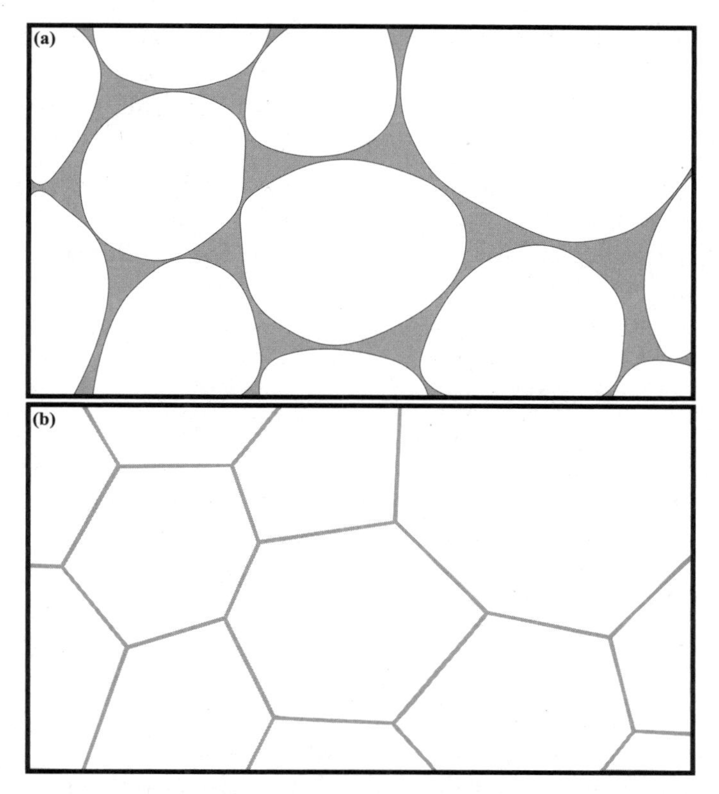

Figure 5.3. The two stages of beer froth. (a) Liquid foam consisting of fairly round bubbles suspended in liquid beer. This wet foam is quite fluid. (b) Dry foam, which results from the beer draining away and from bubble growth. This foam consists of polygonal bubbles and is more solid and stable than wet foam. A head of beer may contain dry foam at the top and wet foam underneath.

More serious work has confirmed that the height of beer foam falls with time in an exponential manner. This matters to brewers because they would like their beer to retain its head: this characteristic is deemed to make the beer more appealing and marketable. So, the problem of *head retention* is not confined to French monarchs. For example, it has been observed, noted, and analyzed to the *n*th degree that grease

---

with Digital Rectal Massage"; Physics—French research explaining why dry spaghetti often breaks into more than two pieces; Chemistry—Spanish research into the changes undergone by cheddar cheese when it moves at ultrasonic velocities.

(lipids, in particular) is responsible for a lot of beer head collapse (see the study by Keusch listed in the bibliography). Thus, if the beer glass contains some residual soap after cleaning, this soap will cause the beer bubbles to burst very quickly and the head will not form or will decay rapidly. If you have been eating greasy food, or are wearing lipstick, as soon as you apply mouth to beer glass you will cause your beer foam to wilt and die. So, to form the foam you must lose the lipids. Another approach: if your draft beer is dispensed with nitrogen gas, then it will retain its head wonderfully well. Canned beer achieves the same creamy, long-lasting head with a nitrogen-dispensing *widget*. More on nitrogen and widgets later.

We can mathematically model the increase and subsequent decrease in beer foam volume. Our model makes predictions about the evolution of beer foam height in the glass from the moment that the beer is poured. It is an incomplete model (there are a lot of parameters) and so should be regarded as a foundation upon which a more detailed model can be built.

In figure 5.4 we see the three components of a newly poured glass of beer. My task is to construct a simple mathematical model that tells us how the volumes of these components change with time. For simplicity I will assume that the glass is straight-sided so that volume is proportional to height (and thus I need not worry about how volume changes with height above the glass bottom).[11] Let us say that, at a given time t, the heights of the dry foam, wet foam, and beer are h_d, h_w, and H, respectively. My model assumptions are shown in the figure. I will assume that, during a short time interval dt, a fraction, $a\,dt$, of wet foam drains away into the liquid beer. Similarly, a fraction, $c\,dt$, generates bubbles that rise into the dry foam. A fraction, $b\,dt$, of the dry foam is assumed to drain away into the wet foam as bubbles burst. Thus, my model assumptions amount to saying that beer is divided into three components and that bubbles are formed and drain away at a uniform rate from one component to a contiguous component. Undoubtedly this is a simplification of the quite complicated physics of

11. In figure 5.4 the glass is not straight-sided, and you can see that a given volume of dry foam will occupy a lesser height than the same volume of liquid beer. This extra complication—converting volume into height—depends upon glass shape; I ignore it because here I want to concentrate solely upon properties of the beer.

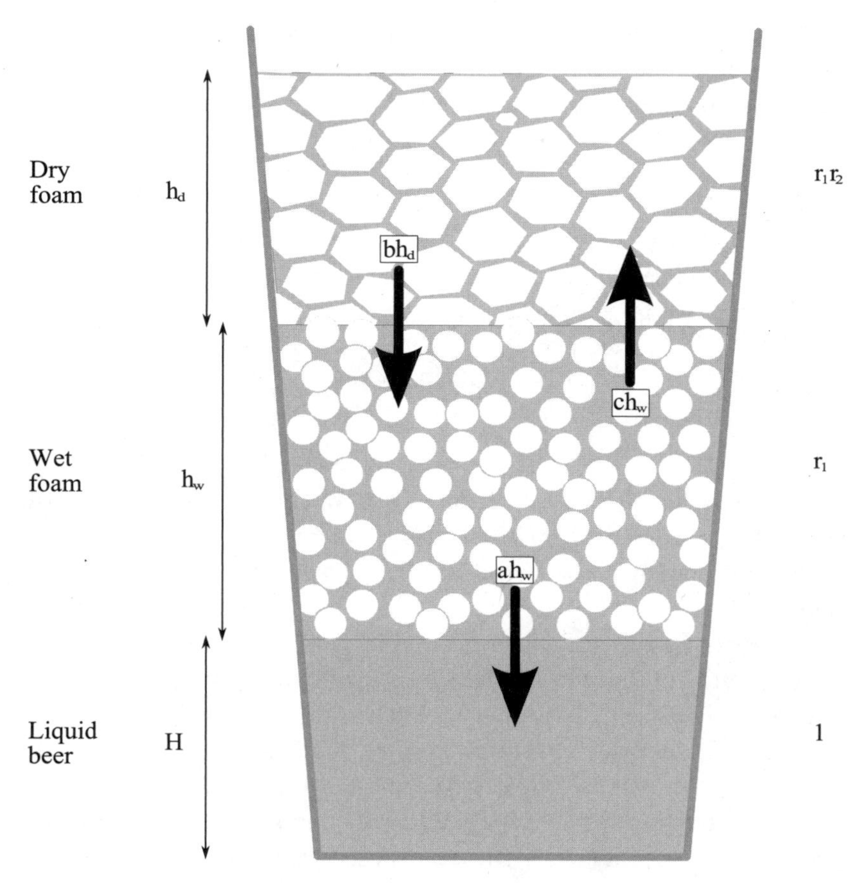

Figure 5.4. From the moment that beer is poured from a bottle into a glass, it begins to separate into three layers: dry foam above wet foam above liquid beer. The volumes of these three components change with time, in a manner that can be described by a simple model. The ratio of relative densities for beer : wet foam : dry foam is $1 : r_1 : r_1r_2$.

beer foam interactions, but as we will see, it accounts for a number of observed features.

The model assumptions lead to the following differential equations that describe the evolution of the beer/foam system over time:

$$\frac{dh_d}{dt} = \frac{c}{r_2} h_w - b\,h_d$$

$$\frac{dh_w}{dt} = -(a + b)h_w + br_2h_d \qquad (5.2)$$

$$\frac{dH}{dt} = ar_1h_w$$

The factors $r_{1,2}$ arise because the density of the three components is not the same. I assume that the density of wet foam is less than the density of beer by a factor r_1 and that the density of dry foam is less than the density of wet foam by a factor r_2 (so that dry foam density is less than beer density by a factor r_1r_2). These additional assumptions are also a simplification of the real world because, for example, there is no reason to suppose that the density of wet foam is constant; if the rate of generation of bubbles is not constant, then wet foam density will vary with time. Nevertheless, the simple model will prove to be illuminating. As I have said elsewhere, one of the skills that a scientist develops, as he or she seeks to understand a particular aspect of nature, is to know when and how to simplify. Skillful simplification can lead to insight, whereas clumsy oversimplification is often misleading.

We can fix r_1, which means that it is not a "free-floating" parameter that can only be determined by observation of beer foam evolution. We have seen that wet foam consists of spherical bubbles that are suspended in liquid. We can reasonably assume that our foam is about as bubbly as it can be (because there is a lot of CO_2 in the beer) without squeezing the bubbles together. (Bubbles squeezed together would be distorted into polyhedral shapes, and this is the characteristic of dry foam. Such foams do not flow like liquids, or like wet foam, because the polyhedral bubbles can't easily slide by each other. You can get a hint of how bubble density affects foam characteristics from fig. 5.5.) From geometrical

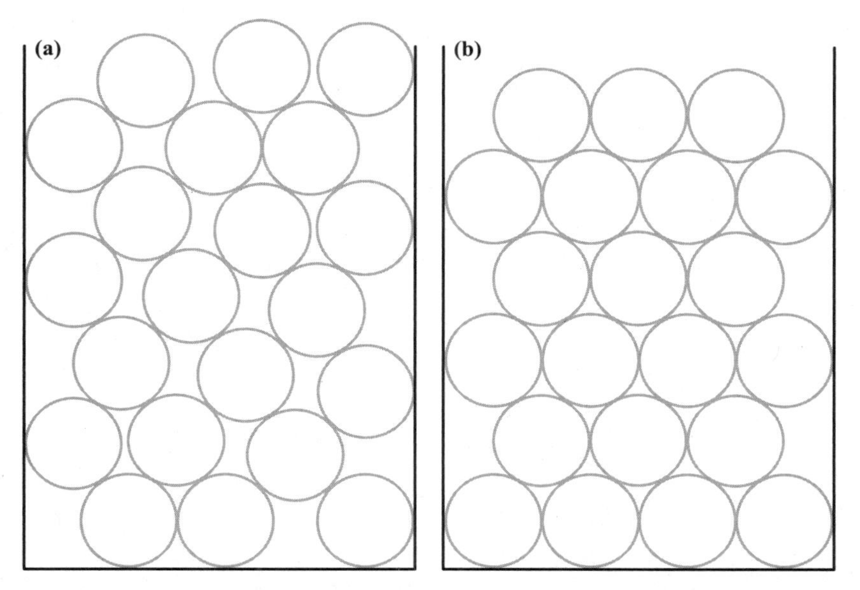

Figure 5.5. We can gain an appreciation of how the density and stiffness of foam changes with bubble density from this two-dimensional illustration. Randomly packed bubbles (a) occupy more space than the same number of regularly packed bubbles (b); thus, the corresponding foam is less dense. It is also less stiff, since the bubbles are able to slide over each other more readily. If the number or size of the bubbles increases, the bubbles crowd against each other and become distorted and eventually polygonal, as seen in figure 5.3.

arguments it is well known that the liquid content of foam needs to be at least 36% in order for the bubbles to remain spherical. Because bubble densities are much less than beer density, we can fix $r_1 = 0.36$.

The equations in (5.2) are solved using standard mathematical techniques.[12] Since no physics is involved, I will skip these techniques and jump straight to the solutions. Even these are a bit of a mess to write down, so instead I will just plot the answer. In figure 5.6 I have reproduced some beer foam data from the brewing literature that show

12. For the math geek inside you who is struggling to get out, I note that the equations in (5.2) are linear coupled ordinary differential equations with constant coefficients and that the solutions are well known to be the sums of exponential functions.

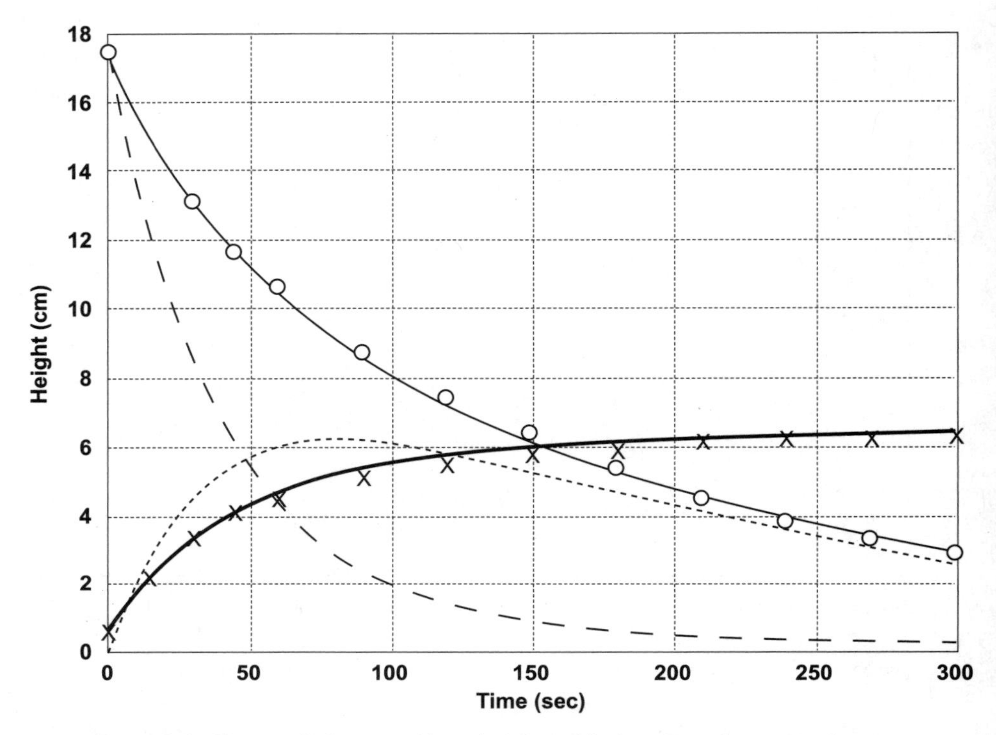

Figure 5.6. Changes in foam and beer heights with time. Data from experiments by Hackbarth (see the bibliography) show changes in foam height for one particular type of beer (open circles). The height of liquid beer is also shown (crosses). My simple model fits the data very well with parameter values of $r_2 = 0.3$, $a = 0.02$, $b = 0.006$, $c = 0.004$. Wet foam height (dashed line) and dry foam height (dotted line) add to give the total foam height (solid line). Liquid beer height is shown by the bold line. This good agreement is obtained by finding the best choice for four parameters; a deeper, more fundamental mathematical model would require fewer such parameters.

for one type of beer how foam height and beer height change with time, measured from the instant that the beer is poured. By varying the four free parameters of my model I can get a pretty good fit. Different beers would evolve differently, so the four parameters would be different for each beer. The same beer poured into a different glass, or poured at a different rate, would also evolve differently and thus would also require different parameters. A more fundamental model would fix these parameters based upon the beer, glass, and pouring characteristics to allow the determination of r_2, a, b, and c before we pour the beer. Such a model would be truly predictive.

My model works backwards: we need to see the data in order to go back and determine what the parameters are. Even so, my model produces some encouraging results which tell us that the basic assumptions must be close to the truth. First, the exponential decay of beer froth emerges naturally from the assumptions. Second, it is easy to verify that the model results are at least qualitatively sensible. Thus, for example, if I pour beer into a glass very quickly, it will fob: the foam level (wet and dry) will increase to a maximum and then fall. This behavior is seen in my model if the parameters are chosen appropriately. On the other hand, beer poured slowly produces foam that starts to decrease from the get-go, as seen in the data (open circles) of figure 5.6. All this variety of behavior emerges from the model as a consequence of different choices made for the four parameters.

It may be possible to improve the model by, for example, measuring the density of wet foam compared to dry foam. Doing so removes the parameter r_2: we are not allowed to vary it at will, just to fit the foam evolution data. It is possible that a more detailed model would find a theoretical connection between the four parameters (for example, $b = r_2 a$). Such developments would increase the predictive power of the model by further reducing the number of independent parameters. However, we leave such ruminations to the more interested reader. The thirstier reader may want to consume the subject of figure 5.7 or similar, after such mathematical exertions. Before doing so, however, please note the bubbles that are forming on the glass and in the head of the beer in figure 5.7. A further study of these bubbles will be the subject of the next two sections.

Disproportionation

The clumsy word "disproportionation" describes a physical process, also known as *Ostwald ripening*, whereby adjacent bubbles containing gas at different pressures tend to pass the gas from one to the other. Naturally, such a process requires a common boundary between the bubbles and so is restricted to the polyhedral (in three dimensions—"polygonal" in two) dry foam, since wet foam consists of bubbles separated by liquid. Because small bubbles contain gas at higher pressure

Figure 5.7. A refreshing brew. But before quaffing it, please note two things about the bubbles: (1) some of the bubbles in the head have grown larger than the others, and (2) bubbles have formed on the side of the glass.

than large bubbles,[13] gas tends to pass from small to large. The de Vries equation describes the time evolution of bubble radius due to disproportionation:

$$r(t) = r_0 \sqrt{1 - \frac{t}{t_0}}. \tag{5.3}$$

So, a small bubble will shrink from its original radius, r_0, to nothing in a time t_0. The collapse time t_0 can be calculated; it depends upon gas pressure and solubility, and upon bubble surface tension and film thickness. The details don't matter to us, with one exception, but the effects of disproportionation do influence the appearance of beer foam.

13. Due to surface tension. Recall how hard it is to blow up a balloon when it is small, compared to blowing it up when it is large. A classics physics demonstration is to connect two balloons (one inflated a lot, and the other a little) via a tube that is fitted with a valve. Turn the valve so that air can pass between the balloons, and see how the small balloon gets smaller and the large balloon gets larger, because pressure and not volume is equalized.

Bubbles that begin their brief lives at almost the same size become very different because of disproportionation. This observation will come as no news to many readers who, like me, have enjoyed staring at beer bubbles with a kind of childlike fascination that has nothing (well, not much) to do with the amount of beer already consumed. Some more scientific musings upon the fractal nature of bubbles have found their way into print (see Saurbrei), and we can readily see how such musings might arise (see fig. 5.8). Real foam cannot be truly fractal, of course, because the patterns do not extend down to infinitesimal scales, but it is possible that over a limited range of length scales, the patterns of bubbles may seem fractal-like.

It is considered esthetically undesirable for beer foam to develop large bubbles.[14] The frothy head looks more uniform, and creamier, when the bubbles are small. Thus, the brewing industry has expended considerable effort to minimize the effects of disproportionation. By slowing down the rate at which small bubbles give up their gas to large bubbles, the unsightly bloated bubbles are banished. The key discovery, which has proved to be so successful that it is very widely applied by many different brewers around the world, is that nitrogen gas in the beer dispenser stops disproportionation in its tracks. That is, beer dispensed from a keg or barrel by nitrogen gas pressure (or canned beer with a nitrogen widget) creates uniform foam consisting of small, long-lasting bubbles. How? Nitrogen gas is much less soluble in beer than is the traditional dispensing gas (which is CO_2, of course, since the beer is deliberately carbonated), and solubility greatly influences the collapse time, t_0, of equation (5.3). Calculations show that t_0 is inversely proportional to the solubility of the gas within the bubble; nitrogen bubbles have a much larger value of t_0 than do CO_2 bubbles.[15] From (5.3) you

14. Why this should be the case I cannot begin to guess. However, if enough tipplers think that big bubbles are bad, the results will show up in the bottom line of brewery sales. So, given the efforts that brewers have expended to influence foam appearance, perhaps we can conclude that people really do prefer small, uniform bubbles in their beer foam.

15. There must be more to the physics of disproportionation than this explanation provides, however, if Bamforth's claims are true (see bibliography). He states that even very small amounts of nitrogen gas—a few parts per million—can significantly hinder disproportionation.

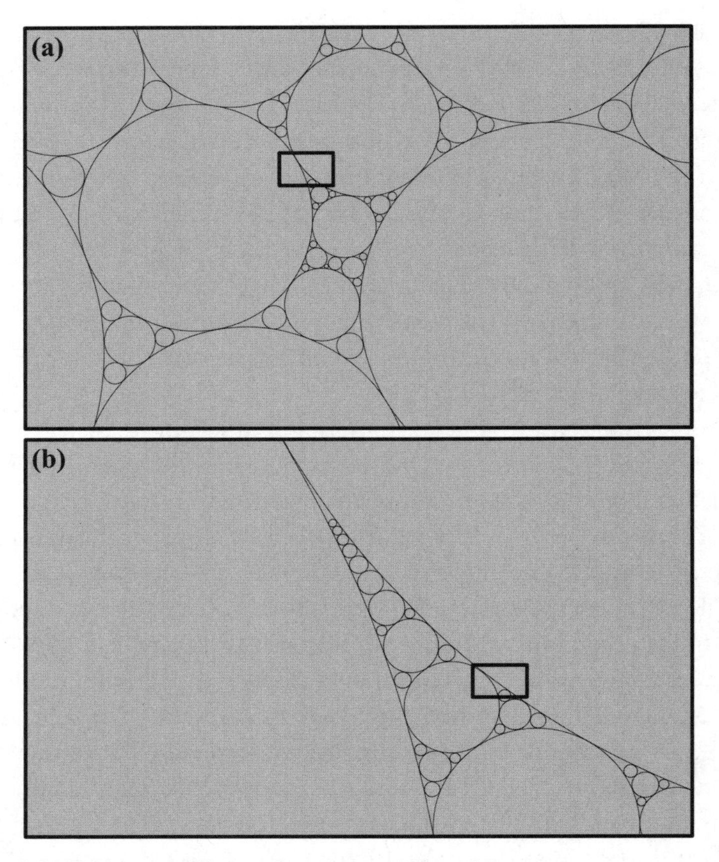

Figure 5.8. Bubbles of different sizes crammed into 2-D space may assume fractal-like patterns. Magnifying the small rectangular section of (a) may yield (b). Magnifying the small rectangle in (b) may yield a very similar pattern. Such self-similarity on scales of different length is a defining characteristic of fractals.

can see that the rate at which small bubbles collapse (and hence the rates at which large adjacent bubbles grow) slows down a lot if the collapse time increases a lot. Hence the rush to nitrogen dispensing.

The nitrogen fixation of the brewing industry[16] is not all good news for beerophiles, however. Nitrogen gas is not a natural part of the brewing process; it has been introduced artificially to solve a perceived image problem. Some beer lovers complain that the flavor of their beer is

16. Technical pun—sorry.

changed as a direct consequence of the use of nitrogen during beer dispensing. Naturally, the replacement of CO_2 by nitrogen in dispensers means less dissolved CO_2 in the beer. The tang, or bite, that CO_2 introduces to beer is thus much reduced.[17] Now, bite is a subjective effect, and many folk do not mourn its passing, but others do. The beer head not only looks different—the reason for introducing nitrogen in the first place—but it tastes different. The hoppy flavors in the head have disappeared, or been much reduced, and the *mouthfeel* (a favorite beer-tasters' word) is different: nitrogen foam feels much creamier. The effects of nitrogen on the beer as a whole can be summarized by saying that the beer tastes softer and milder. For a bad beer, these effects may be a good thing, but a good beer can be made more bland and ordinary. The traditional craft brewer does not brew beer to be dispensed with nitrogen, and so this type of beer (unlike MB beer) does not benefit from nitrogen dispensing.

Before moving on, I will insert a widget. Widgets are an invention of the 1980s that won an award for industrial innovation. Widgets are responsible for providing you with canned Lyte beer (and also more acceptable beers that are dispensed from cans) that displays a great-looking creamy, foaming, frothy head. Widgets bring nitrogen dispensers into the beer can, and they work as follows. A plastic sphere—the widget—hollow and with a small hole, is placed in the can with the beer. Nitrogen is injected into the can, under pressure, and the can is then sealed. Some of the beer and nitrogen is pushed into the widget through the small aperture. When you pull the tab, pressure is released and beer plus nitrogen gas spurts out of the widget and into the beer. The resultant mechanical agitation (plus that of pouring) brings most of the nitrogen out in the form of small bubbles, so what emerges in your glass is creamy, wet foam. *Voila*. The widget is effectively acting as a beer tap restrictor, which serves the same function for draft beers. These restrictors (there are many variants) can be adjusted so as to cause the beer to foam in the desired manner. Widgets are inexpensive restrictors, and they perform well because they are used in conjunction with nitrogen.

17. Guinness attempts to have the best of both worlds by dispensing its famous brew with a mixture of 75% nitrogen and 25% CO_2.

Drainage

Readers who are knowledgeable experts in our subject will be wondering how I can get so far into a chapter about beer foam and beer bubbles without mentioning *surfactants*. A surfactant (the word is a contraction of "*surf*ace *act*ing *agent*") is usually a protein molecule—though starches and sugars may act as surfactants—that is present to a greater or lesser extent in beers, but always more than in sparkling wines and other carbonated beverages. It is the surfactants that are responsible for the very different behavior of beer bubbles and beer foam, compared with other fizzy drinks. Surfactants effectively make beer sticky, like toffee, compared with carbonated mineral water or champagne. Beer bubbles become coated with surfactant molecules. This coating increases the bubbles' hydrodynamic drag as they plow their way through the beer. So, beer bubbles rise more slowly than do champagne bubbles. (They also fall, as we will see.) The surfactant coating acts to retard bubble coalescence—bubbles burst less easily—and it influences disproportionation, which is why I might have mentioned surfactants earlier. Much of the explanation for surfactant behavior lies in the realm of chemistry and so is somewhat outside of my chosen area—enlightening you about beer physics—but the effects of surfactants are certainly physical. We have all observed the longevity of beer bubbles, compared with those in champagne or cola.

In fact there is a complicated interaction between surfactants that attach themselves to beer bubbles, so that both chemistry and physics are required to explain the total effect. This area is still being researched because it is, as you might imagine, complicated. It seems to be the case that as liquid drains from the wet foam, leaving behind increasingly dry foam with thinning walls, different surfactants interact with each other to produce local variations in viscosity. Now, viscosity is a key parameter that influences, among other things, drainage. So here we have a complicated interaction of chemical and physical effects that makes beer foam more interesting to physicists than champagne bubbles.[18] The drainage of beer from wet foam is complicated enough with-

18. Of course, the fact that beer is more within the physicist's budget is an irrelevance, you understand.

out having to worry about interacting components, though in fact physicists have long known how to calculate the drainage rate for two-dimensional foams. The two-dimensional drainage rate, often quoted in the technical literature, is given in terms of beer density and viscosity, and in terms of bubble size and wall thickness. But of course beer is three-dimensional.[19] Recently, mathematicians who favor beer (a large subset of the whole) have joined the real world by calculating for the first time the way in which 3-D bubbles coalesce. I direct those of you who are interested in more details to the article by MacPherson and Srolovitz.

Real (3-D) beer drains from top to bottom, of course, and so real beer foam is of variable consistency and character. It may be dry and stiff at the top, with thin-walled polyhedral bubbles, and be wet and fluid at the bottom, with spherical bubbles. Such variability further complicates foam analysis. To this end, a NASA publication (Durian and Zimmerli) advocates observations of foam in space, where an absence of the effects of gravity makes foam more uniform. I don't think NASA is proposing that astronauts take a six-pack of beer into space with them,[20] but it would be highly entertaining to see them try to pour beer from a pressurized can into a glass at zero gravity.

STANDING, RISING, AND FALLING BUBBLES—AND ANTIBUBBLES

We have poured our beer, and now it is sitting in the glass, awaiting consumption. The wet foam has transmuted into dry foam and liquid beer, with a clear dividing line between the two. Several minutes after pouring the dry foam has diminished, but it is still there. Bubbles are forming at nucleation sites on the glass (see fig. 5.7); these occasionally detach and rise to join their friends among the foam. Different beers generate different bubbles, and the bubbles rise at different rates. We have seen that surfactants coat beer bubbles and increase the drag force affecting these bubbles; the drag force, in turn, counteracts the upward buoyancy force. Another factor that influences the rate at which bub-

19. 2-D beer would probably taste quite flat, don't you think?

20. Could we describe taking beer into space as "elevating the spirits"?

bles rise through the beer is bubble size. Bubble speed through a stationary fluid is given by Stokes' equation:

$$v = \frac{2\,\rho g}{9\eta}\, r^2, \tag{5.4}$$

where η is beer viscosity and r is bubble radius.[21] It makes sense that bigger bubbles should rise faster than smaller ones because the buoyancy force increases with bubble volume (Archimedes' principle). Equation (5.4) applies only when the liquid containing the bubble is not moving—or, more precisely, it gives us the speed of the bubble relative to the liquid.

The point is that, for our glass of beer containing bubbles, the beer is not stationary within the glass. Circulation patterns are set up as a consequence of bubble movement. These circulation patterns look much like the circulation of hot wort, sketched in figure 4.5. The reason for these patterns, in this case, is as follows. Bubbles throughout the glass rise to the surface, and in rising they drag some of the beer with them. But the entire glassful cannot rise up: the net speed must be zero, since your pint of beer does not leap skyward unaided, except perhaps very late on a Saturday night when you have had a few too many. Beer rises in the center of the glass, where resistance due to the side walls of the glass is a minimum. Beer descends down the outside, next to the glass, to fill in the void left by rising beer. Hence the pattern of figure 4.5. This circulation pattern affects bubble speed, as seen by the external observer. Bubbles near the outside rise more slowly than bubbles near the center. In fact, for very small bubbles (those whose natural speed of ascent is very slow) the descending speed of the beer exceeds the ascending speed of the bubbles, and the net result is that very small bubbles near the side of the glass may actually *fall*. Guinness drinkers will be familiar with this phenomenon—it is a fascinating and quite beautiful sight, this choreography of dancing bubbles. The critical di-

21. Equation (5.4) applies to the "terminal speed" of the bubble. Recall that skydivers accelerate through the air until drag force (which increases with speed) balances the force of gravity, at which point the skydiver falls with constant speed—his terminal speed. (Terminal in another sense if his chute fails, I suppose.) For beer, the terminal speed of bubbles is a leisurely 0.1–0.5 cm s^{-1}.

ameter for bubbles to fall is about four thousandths of an inch (a tenth of a millimeter). Smaller bubbles fall, larger ones rise.

Beer bubble behavior is more complicated still, because of the action of surfactants. You may have noticed that bubbles grow as they rise up through the beer. A bubble at the top of your pint may be twice as big as when it first detached from its nucleation site and began to rise. "Aha," you say, "surely that is because the pressure of the beer is greater at the bottom of the glass than at the top." In fact, no; the glass would have to be 240 feet tall. That is, if pressure caused bubbles to compress, so that a bubble at the bottom was half the size of a bubble at the top, then the difference in height between top and bottom would have to be 240 feet. A tall order, indeed. The true reason why bubbles grow has been unearthed only within the last 20 years, by Zare and co-workers (see Shafer and Zare). Surfactants not only coat the bubble surface and slow the bubble's rise to the top, but they also cause the bubble to further accumulate dissolved CO_2 in the beer. The bubble, in other words, becomes self-nucleating, like a snowball gathering mass as it rolls downhill. Needless to say, this behavior of bubbles is not seen in carbonated water or other fizzy drinks, since they lack surfactants.

Zare has further calculated that beer bubbles act strangely. When they are smaller than a third of a millimeter they are spherical, but when larger than that they are ellipsoidal in shape. In both cases they accelerate through the beer rather than move at a constant speed.[22] Furthermore, in very tall beer containers (such as the "yard of ale" discussed in chapter 6) bubbles can grow as large as a millimeter; when this happens, calculations show, the behavior of a bubble becomes truly weird. Its radius begins to oscillate as the bubble rises, and its movement is no longer a straight line: it zigzags or spirals, as if drunk.

And then there are *antibubbles*. These strange beasts are exactly the opposite of bubbles. Whereas a bubble (say a soap bubble) is a hollow sphere—a spherical shell—of liquid surrounded (inside and out) by gas, an antibubble is a "hollow" sphere of gas surrounded (inside and out) by liquid. Antibubbles have been created in Belgian beer. I kid you

22. Consequently, we must regard the Stokes equation (5.4) as only approximately true for beer bubbles. Recall that this equation assumes the bubble to be of fixed radius.

not. If you happen to be one of those people who feel unfulfilled or inadequate due to a lack of knowledge of antibubble phenomena, then fear not: help is at hand. It turns out that antibubbles do not occur in pure liquids such as water or alcohol. They require surfactants. A liquid containing surfactants (Belgian beer, for example) is poured into a glass of the same liquid. The poured liquid drags air with it into the glass, coating droplets and providing a gaseous barrier between the poured droplets and the liquid in the glass—antibubbles. For Belgian beer, the antibubbles that have been created are large, and they persist for up to two minutes. Without surfactants the dragged air does not coat the poured droplets, which simply mix with the liquid in the glass. It will make your day to learn that physicists have developed a successful mathematical model of antibubble formation; see the David Reid article "Scientists Create Antibubbles," which shows you how to create your own antibubbles.

Six

Fluid Flow

Come, my lad, and drink some beer.
—Samuel Johnson (18 September 1777)

If God had intended us to drink beer,
He would have given us stomachs.
—David Daye

CIRCULATION

Thus far we have seen circulation of fluid in two guises: wort circulating (or not, if it is thick enough) in the homebrewer's fermentation bucket and beer circulating in the glass. From the point of view of the brewing industry, however, these examples are small beer compared with the problems of beer distribution that a large brewer has to solve. In this chapter I will look at how beer gets from the fermenter in a brewery to the stomach in a beer drinker. At that point we will end the chapter: I have no interest in telling you about the subsequent career of your beer. The fable in chapter 1 suggested that Lyte beer short-circuited the brewery entirely: it passed from the stomach to the bladder and from the bladder back into the beer barrel, perhaps pausing only to be chilled and carbonated before beginning the cycle again. In fact, the distribution of beer from brewery to liquor store, or bar, café, or pub is a serious issue that helped shape the large and small breweries that we see today. We will follow the journey of a pint of beer, beginning where we left off in chapter 2, with beer that has matured and is ready to bottle. I will assume that this pint was produced in a commercial brewery, not in a homebrewer's kitchen or storeroom. We will see our pint barreled, bottled, or canned, then distributed to a pub cellar, then dispensed into a glass, and finally glugged down by a thirsty beer quaffer. On this journey we will encounter some interesting engineering physics and mathematics.

PACKAGING (COLD ROOM TO CONTAINER)

From the dawn of beer, brewers have been faced with the problem of how to get their product from the brewery to the customer. In medieval Europe (where, you may recall from chapter 1, monasteries dominated brewing in many countries for several centuries), the monks solved this problem simply by obliging the customer to come to them. In later and more competitive times, the onus on supplying beer fell to the brewers, and they needed to find a method of transporting their precious brew without spilling or spoiling it. The earliest solution for mass transportation of liquids was the barrel; for hundreds of years barrels were the main method of conveying beer from a brewery to the taverns of Europe.

Making wooden barrels is a skilled craft and developed into a significant trade. Coopers would shape curved wooden staves, carefully beveled and tapered, so that they fit together closely. These staves were held in place with iron hoops. The barrel interior would be *pitched* with wax or resin to separate the beer from the wood (see fig. 6.1). A finished barrel was watertight, strong, and easy to use. It could be rolled, toted on wheels, or floated on water. So ubiquitous were barrels (which were used for transporting many different foodstuffs and other items of value) that the form of barrels—casks, hogsheads, kegs, tuns, etc.—became specialized by shape and size, depending upon the purpose. "Cooper" became a surname as well as a trade.

By the early sixteenth century almost all beer was transported by barrels. In England, to promote fair trade it was made illegal for brewers to make their own barrels, and a standard size (36 gallons) emerged. After being placed on its side or end in the tavern cellar, a barrel was pierced by driving a spigot through a soft plug in the barrel end, as shown in figures 6.2 and 6.3. A second, smaller vent hole was opened at the top to prevent a vacuum developing as beer was poured from the

Figure 6.1. "Pitching" a barrel of beer—i.e., coating the interior with an inert waterproof seal that separates the beer from the wood of the barrel. Photo courtesy of the Bavarian Brewers Federation, Munich, Germany.

spigot tap. The vent let in air, which meant that the beer had to be sold quickly because it would soon spoil. And even before souring, the beer would lose carbonation. By the end of the nineteenth century, cellarmen had learned to connect the vent to a carbon dioxide source so that no air entered the barrel as it emptied.[1] This, and the cool cellar temperature, greatly increased the life of beer in a partially emptied barrel.

Nowadays more hygienic—if less interesting—stainless steel or aluminum kegs (see fig. 6.4) replace wooden barrels, except among diehard traditionalists. Carbonated and perhaps pasteurized beer can keep for up to three months in a keg, but only for one month at most in a traditional cask. Today the fraction of all beer that is transported in wooden barrels or steel kegs is falling (it is down to 10% of sales in the United States, from 70% in the 1930s) and has been falling for a century due to the rise in popularity of beer bottles and, later, beer cans.

Glass containers are as old as beer, and glass bottles have been used as beer containers in Europe since the sixteenth century or earlier. By 1695 in England there were three million beer bottles in circulation. Even in the eighteenth century, though, these bottles were expensive. They were also irregular because each was handblown. By the nine-

1. By 1900, 75% of U.S. beer outlets used CO_2 in this manner.

Figure 6.2. A barrel must be provided with a bung hole for a spigot tap. Photo courtesy of the Bavarian Brewers Federation, Munich, Germany.

Figure 6.3. Tapping a barrel can be a messy business. Photo courtesy of the Bavarian Brewers Federation, Munich, Germany.

Figure 6.4. The ubiquitous beer keg. Photo courtesy of the Bavarian Brewers Federation, Munich, Germany.

teenth century bottle manufacturers had mastered the technology of mass-producing glass bottles of a standard size and shape, and this is the period when bottles came to dominate as beer containers. At the beginning of the nineteenth century the United States exported a million gallons of beer, and 15% of this was in bottles. As the century wore on, both the volume of beer transported, and the fraction that was transported in bottles, increased. Bottles were more convenient for smaller retail outlets, and for customers to take home. In 1800 beer bottles were sealed with cork stoppers (and wires, to prevent the carbonated beer from pushing out the cork), but in 1892 the crimped metal crown was invented. It was a less expensive, faster, and more effective seal than the cork stopper and has been the standard seal for beer bottles for over a century.

A decade before the crown seal appeared, the problem of beer *skunking* in bottles was solved in Germany. One trouble with transparent glass bottles is that they let in light. Light of certain colors can interact with some hop products within the beer to quickly produce flavors that are seriously off. The descriptive term "skunked" characterizes the resulting beer. It was found that brown bottles do not transmit the

Figure 6.5. A modern bottling plant. Photos courtesy of the Bavarian Brewers Federation, Munich, Germany.

offending colors of light, and so beer bottles became brown to prevent beer from skunking.[2]

In the 1880s Adolphus Busch, and his great rival Frederick Pabst (two of the great American beer barons) realized that bottles were the way of the future, and they invested in bottling plants (along with bottle factories and washing equipment). By 1900 bottled beer was hugely profitable; the breweries, their distribution networks, and the number of beer bottles, all grew. Busch's bottling plant in St. Louis had a capacity of 700,000 bottles. The big breweries of this period were becoming vertically integrated behemoths, controlling all aspects of beer production and distribution. Beer bottles were poised to take over the world (see fig. 6.5).

Today there are dedicated bottle-feeding machines that present clean beer bottles to the bottle-filler (see fig. 6.6). Flash pasteurization is used in some countries such as England to preserve beer before bottling.[3] The chilled beer (from the brewery's cold room, where it is stored) is then artificially carbonated in the bottle and quickly crown-capped. A large modern bottling plant can fill 60,000 bottles an hour. The bottle exterior is dried to remove condensation before the label is attached (there are at least two different systems for attaching labels). Finally, the bottles are packed into cases, either manually or by machine. My brief description of the bottling process covers the basic process in a large brewery; there are variants, especially for small-scale craft brewers.

Beer cans were invented in 1935, and they proved to be a lighter, less expensive, and more robust alternative to bottles. By 1969 more beer was sold in cans than in bottles, and today 60% of American beer is sold in cans. Whether beer is canned or bottled, its packaging is the most

2. Some beer bottles today are green. This is a gimmick or a marketing ploy and has nothing to do with preventing the beer from skunking. Beer in green bottles will skunk just as much as beer in clear glass bottles. A recent innovation by some of the larger brewers is to modify the hop extract they use in brewing, so that even in clear bottles the beer does not skunk. My green plastic bottles (discussed in chapter 2) when filled with homebrew are stored in dark cupboards so that light cannot interfere with my precious beer.

3. In flash pasteurization the beer is raised to over 70°C for 15–30 seconds. Because English beer is stored at cellar temperature, it goes sour more quickly than does American lager beer, which is stored and served at much lower temperatures. Hence, in general, English brewers need to pasteurize whereas U.S. brewers do not.

Figure 6.6. Beer bottles being filled and capped. Photo courtesy of the Bavarian Brewers Federation, Munich, Germany.

complex, most labor-intensive, and most costly part of the entire commercial brewing process. Astoundingly perhaps, fully 60% of the capital expenditure for commercial beer production and distribution is due to packaging; only 20% is due to brewing the beer.

Unfortunately, the packaging process has a deleterious effect upon beer quality. It can interfere with the freshness and flavor of the beer and introduce flavors of its own, unintended by the brewer. Pasteurization cooks the beer, knocking out flavor. The widely used alternative of sterile filtration removes yeast cells as well as bacterial cells (which are much smaller)—but it also removes some flavor. Of course, both pasteurization and sterile filtration are applied in order to help preserve the beer for extended periods of time, as it is packaged and distributed. So, inevitably, beer that is packaged and distributed will taste less fresh and flavorful than draft beer that is served near the brewery. For these reasons smaller brewers, with a local distribution, have an advantage over larger brewers when it comes to the freshness of their product. Chilling and filtration or pasteurization may preserve the beer, but not the freshness. A beer brewed in Milwaukee and drunk in London cannot be as young and full-flavored as a local London beer that has not been subjected to the rigors attendant on long-distance transportation. Put more succinctly, beer that has come a long way often tastes jet-lagged. A further example: in addition to having been filtered, beer that is packaged for transportation is usually carbonated artificially: this is more reliable than the traditional method of priming the beer so that it conditions naturally in the barrel or bottle, and it leaves no sediment. But artificial carbonation can markedly change the character of a beer. Many of you beerophiles will have enjoyed a cask-conditioned draft brew and subsequently been disappointed by the canned or bottled version. There is no substitute for naturally brewed and conditioned draft beer, unfiltered and drunk fresh.

DISTRIBUTING (BREWERY TO CELLAR)

Our pint of beer from a large brewery has been packaged and is now being sent out to retail outlets. The traditional outlet was the tavern cellar, but these days it is more likely to go to a wholesaler. The beer distribution system in the United States was originally like that in England, but the system changed drastically following the repeal of Prohibition. In England, breweries own *public houses,* or pubs (an old example is shown in fig. 6.7), where the beer is dispensed to thirsty patrons. Large breweries own thousands of pubs, and generally these pubs serve only the beers made by the brewer who owns them. This was also true in the United States prior to Prohibition: we saw earlier that large breweries became vertically integrated and that by the end of the nineteenth century they were distributing their beer nationwide to their own bars and saloons. U.S. lawmakers, however, considered the practice of breweries' owning pubs to be uncompetitive and after Prohibition decided to make a clean sweep of past practices. When alcohol consumption was once again legalized in 1933, federal and state lawmakers established the current three-tier system of beer distribution, whereby all brewery beer has to pass through a middleman[4] to ensure that pro-

4. Ahem, wipe that smile off your face—you know what I mean. I am certainly not suggesting that U.S. law requires beer to be filtered through wholesalers' kidneys prior to public consumption.

Figure 6.7. An old English pub. This one is in the southern seaside town of Hastings. Thanks to Mike for this image.

ducers cannot control the retailers and limit customers' choices to the producers' own products. So, brewers and importers sell their products to wholesalers, who then sell the beer to retail outlets—bars, restaurant and hotel chains, and convenience stores.

In England, the old system seems not to be uncompetitive. The *tied houses* of one brewery (those pubs that are owned and operated by the brewery) share a town or village with the tied houses of another brewery, and also with *free houses* (pubs that are not owned by any brewery and can sell any beer they choose)—often on the same street. Thus, a consumer who is dissatisfied with one beer or pub has plenty of choice within walking distance. What works in crowded England may not work in sprawling rural America, and it is easy to imagine that consumer choice was greatly limited in the early-twentieth-century small-town USA where only one brewery was represented in a remote location or a hard-to-reach mining camp. The current three-tier system is a patchwork, since much of the legislation was left to individual states. Thus, regional variations in beer distribution laws can be significant. Given the variability of state laws, it is perhaps not surprising that the modern revival of craft brewing is far from uniform, geographically speaking.[5] There has been much—too much—said about the merits and demerits of the three-tier system, and I do not intend to say anything more about it here. Instead, I will construct a mathematical model that will serve to illustrate something of the complexity that the distribution of beer entails for the brewery accountant.

Within the business education sector there is a well-known game called the *beer distribution game* (it is available on the Internet, for those of you who are interested). The beer distribution game is a simulation board game that was invented at MIT's Sloan School of Management to demonstrate to economics students the principles of supply chain management. Perhaps the worthy professors recognized the difficulties encountered by brewers in distributing a perishable product over thousands of miles. They saw that the effects of technology (the invention of refrigeration, say) changed the equation totally, and they saw how marginal differences in production costs or pricing policy could have drastic effects on brewery profitability. Or maybe the simple truth is that beer is never far away from the minds of most academics. I don't know which is the case, but the point is made: beer distribution serves to illustrate a number of factors pertaining to supply chain management.

My toy model—toy because, for ease of presentation, I greatly simplify the problem—will show how profitability is sensitively dependent upon certain production and distribution costs and less sensitively dependent upon others. It shows that the size of a profitable distribution network is governed by production and distribution efficiency, and it shows that you don't need a degree from MIT to drink beer. Well, OK, so you already knew the last bit.

Consider the small, isolated community of Grid City, in Square State, USA, say a century ago. You are the CEO of the Flea Pee Brewing Company. Flea Pee will, in the following decades, be one of hundreds of small regional brewing enterprises that are swallowed up by a MB. For now, however, you are less concerned with out-of-state competition than with your local beer distribution network, which is illustrated in

5. For example, nowadays some states permit craft brewers to distribute their own beer, whereas others insist that they must stick to the three-tier system.

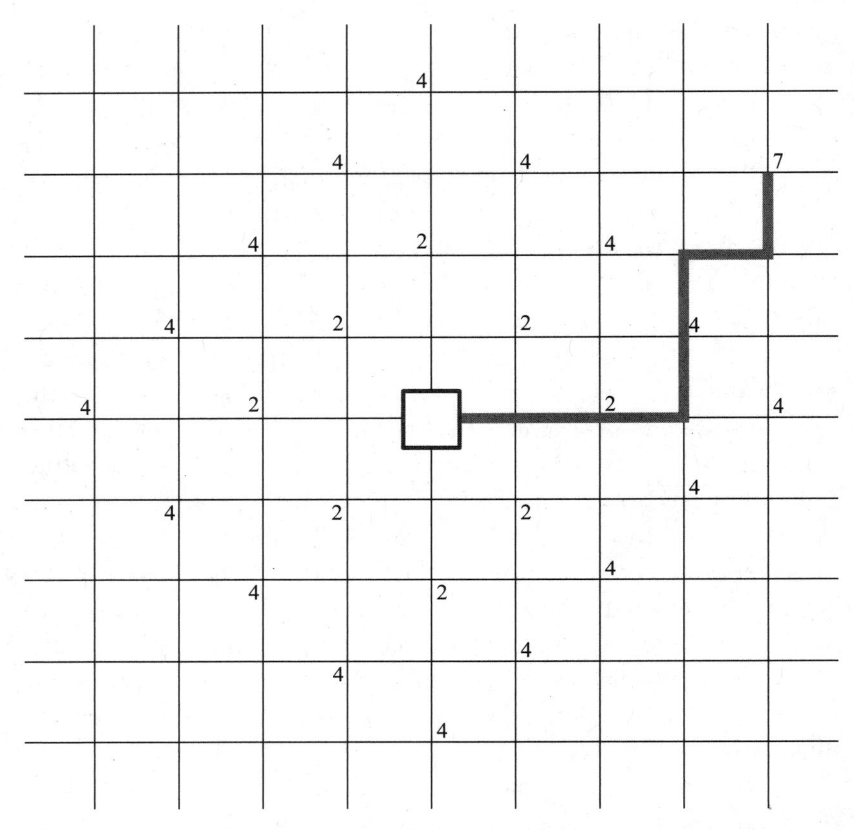

Figure 6.8. The Flea Pee Brewery is located at the center of this grid, and at each grid intersection there is a bar. So, for example, there are 8 bars at two blocks' distance from the brewery, 16 at four blocks' distance, and 28 at seven blocks' distance. The simplest distribution model assumes that the beer is delivered from the brewery to each bar individually. One possible route for a bar at seven blocks' distance is shown.

figure 6.8. Note that your retail outlets (small bars at the corner of each block) are provided with their beer directly from the brewery. So, for example, the bar at seven blocks' distance gets its beer via, say, the route shown in figure 6.8. The cart that delivers the beer barrels does not replenish any of the other bars on the way—these bars have to arrange their own delivery. This system is not very efficient, but it is convenient because you do not have to coordinate delivery between bars. Each bar simply requests more beer when it needs it.

The delivery network shown in figure 6.8 applies for other breweries. Your Canadian cousin Kay Beck runs the Brimming Potty Brewing Company (destined to be taken over by a Canadian MB). Brimming Potty beer is distributed to Brimming Potty taverns that are located at the corner of rural route intersections; the distance across each block is now 20 miles and so the beer distribution costs are different, but the principle is the same. Another cousin of yours, Dan Under, runs another brewery in Australia. The Affluent Effluent Brewing Company (which in time will grow to be one of the largest MBs in Australia[6]) delivers to pubs that are also 20 miles apart, but Dan's distribution problems are not the same as Kay's. Dan's brewery is located in the middle of a desert, and so the cost of raw materials for his beer is higher.

So, you and your far-flung cousins have three similar products with similar distribution systems, but as we will now see, the economics of each of the three breweries will be very different. Let us say that the cost of producing and packaging one barrel of your beer is c dollars per barrel and that the price you charge each outlet is p dollars per barrel. The cost of distributing each barrel depends upon distance, and for you in Grid City it is d dollars per barrel per block traveled. The profit that you make when you sell a barrel of Flea Pee beer to a bar that is n blocks away is $p_n = p - c - dn$ dollars. So far, so good. Let us say that you produce B barrels of beer per week per pub, on average, and that you deliver to bars as far away as N blocks. We see from figure 6.8 that there are $4n$ bars that are at distance n. Therefore, the total weekly profit you make is

$$P = B \sum_{n=1}^{N} 4n\, p_n. \qquad (6.1)$$

6. You may be interested to learn that in 1947 Affluent Effluent will buy out the Dingo Sewerage Brewing Co., and in 1959 will merge with the Asian Guano conglomerate before amalgamating in 1968 with Trans American Cistern Overflow to form one of the world's largest brewery empires, name withheld. This multinational corporation will invent Lyte beer in 1973 and will supply the Chuggaluggers Bar in Milwaukee with same. Small world, ain't it?

In words:

The weekly profit is the number of barrels produced per week per pub, multiplied by the profit per barrel, added up over all the pubs you provide.

We can calculate this profit. For simplicity let us assume that N is large—you are a growing company—in which case the profit is approximately

$$P \approx 2BN^2\left(p - c - \frac{2}{3}\,dN\right). \qquad (6.2)$$

This equation is interesting: it tells us that you had better be careful and not expand too much because if N gets too big, your profit turns into a loss. In fact, there is an optimum size of your Flea Pee brewing company, given your pricing and costs. It is $N_{opt} = (p - c)/d$, which gives you a maximum weekly profit of

$$P_{max} \approx \frac{2}{3}\,B\,\frac{(p - c)^3}{d^2}\,. \qquad (6.3)$$

Let us say that you have expanded your operation to the extent that the number of pubs you supply is optimum, and so your profits are maximized. Now see what happens if circumstances change. If you persuade your customers to drink more Flea Pee beer, so that B increases by 1%, then your profit increases by 1%. If, on the other hand, your distribution costs fall by 1%, your profit increases by 2% so long as you expand the number of bars, N, that you supply by 1%. If the net profit, $p - c$, of a barrel of Flea Pee is increased by 1%, then your total profits increase by 3%, again assuming that you increase N by 1%. So, we see how the brewery profitability depends differently upon different parameters. You may have noted that my toy model makes a simplifying assumption (in fact it makes many): all the parameters are independent. I assume that, for example, if you raise prices, p, this will not influence B, the number of barrels sold per week to a pub. Almost certainly this is not the case in a competitive environment where customers can go elsewhere for cheaper beer.

Equations (6.2) and (6.3) apply for your cousins, too, but with different values for the parameters because of their different circumstances. Thus, both Kay and Dan suffer higher distribution costs, *d*, because the distance between their bars is 20 miles. The cost, *c*, for Dan to produce a barrel of Affluent Effluent is greater than the cost to Kay of producing a barrel of her Brimming Potty beer because Dan's raw materials cost more. So, Dan and Kay will have different optimum sizes for their breweries, and these will in general be different from yours. They will have different profits and different sensitivities to changes in the market and changes in raw material and distribution costs, even though the same equation applies for all three breweries.[7]

The real situation is much more complicated, of course, because there are many more parameters in the real world than I have put into my toy model, because these parameters are not all independent of each other,[8] and because competition with other breweries will rock the boat. The toy model serves, however, to indicate something of the complexity of producing and distributing a perishable product in a changing world. One simplification that I will briefly address is the method of supplying retail outlets. I obliged you to send barrels of beer to each bar individually, without reference to the needs of other bars. This is, of course, inefficient. If two adjacent bars both need new supplies of beer simultaneously, it makes sense for you to use the same cart to take beer barrels to both bars, rather than send two carts. The route that you should take, and how many carts you should use, is a delicate question. I will again provide you with only a brief flavor of the problem here.

You have yet another cousin, this one in England. About a century ago John Bull (why not?) ran the Colorful Animal Brewing Company, and he supplied his beer to 12 pubs, as shown in figure 6.9. His beer was

7. You have two more cousins in Chicago. One runs the Goose Eye Brewery, and the other runs the Land Brewery. They brew excellent craft beer and distribute it locally, so they do not have the distribution problems that you have. Their beer is delivered and dispensed as fresh draft beer, even when they merge to form a single, larger microbrewery (Goose Eye Land). Small is beautiful.

8. I have indicated already how *B* might depend upon *p*. Another example: if your brewery expands so that distribution is further afield, you may have to take extra measures to ensure that your beer does not spoil (increase refrigeration, for example). So *d* depends upon *N*, which in turn depends upon the other parameters.

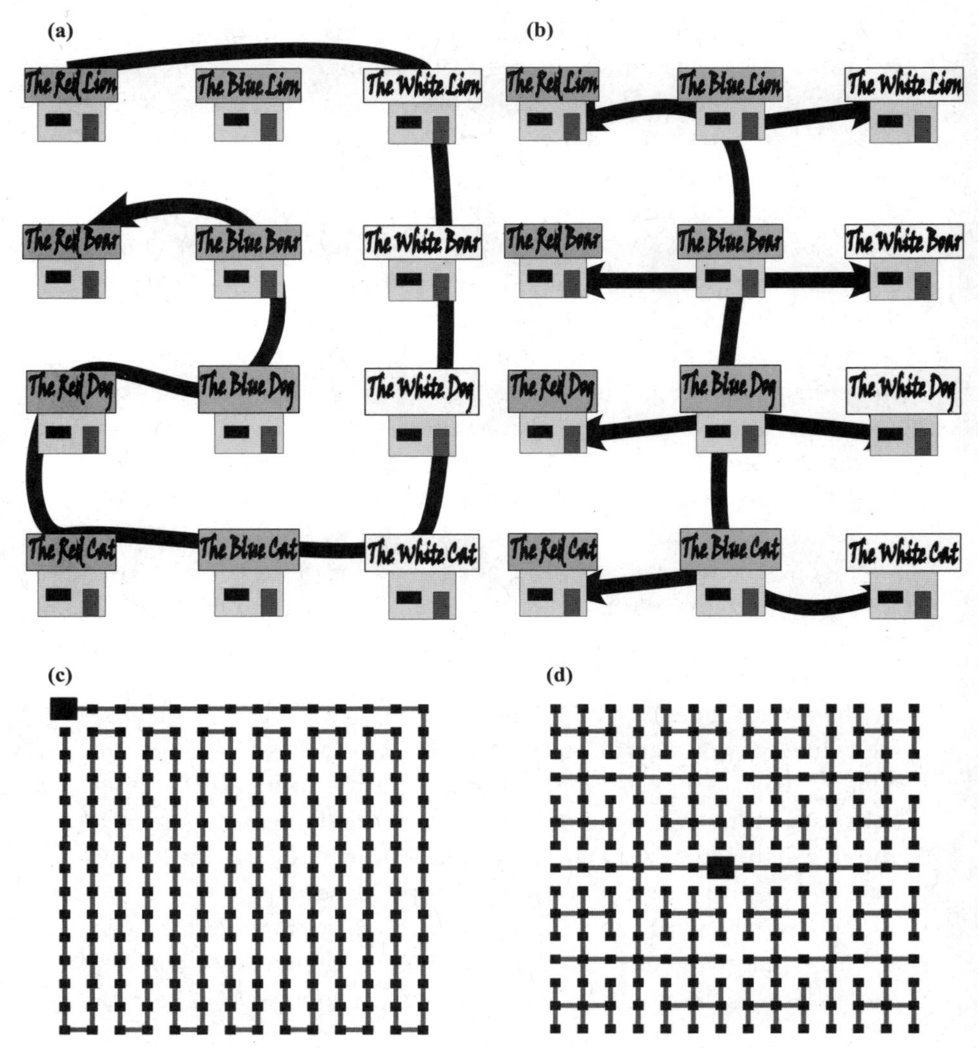

Figure 6.9. A pub crawl. Here I assume that beer delivered to one pub may be redistributed to several others. (a) To supply 11 other pubs with beer, using only one dray starting from the Red Lion, costs 66 bbl-miles. If the center of distribution is moved, and more than one dray is employed, then the cost can be significantly reduced. (b) The cost of distributing beer to all the other pubs starting from the Blue Lion, using the route shown, is 26 bbl-miles. Starting from the Blue Boar it is 20 bbl-miles. (c) For this large network, the distribution cost is 21,945 bbl-miles. (d) But for this network, with the distribution point centralized, the cost is only 2,303 bbl-miles.

carried in barrels loaded onto a traditional dray, much like those of two present-day English breweries shown in figure 6.10. (The old beer drays are no longer economical, of course, but are much loved by brewing companies and by the public. See fig. 6.11.) During the first years of his brewery John delivered barrels of beer to the Red Lion pub, which had a large storage cellar, and then used his single dray to distribute barrels to the other pubs. Despite being in England, these pubs were also arranged on a grid, say one mile apart. So, you can see from figure 6.9 that John's beer distribution cost to resupply his chain of pubs was 66 bbl-miles. How come? Let us say that he puts 11 bbls on his dray at the Red Lion and follows the route shown in figure 6.9(a). He travels one mile to the Blue Lion and drops off a barrel. That part of the journey cost 11 bbl-miles. He moves on to the White Lion and drops off a barrel. This second leg cost 10 bbl-miles. And so on. The total cost, you may like to check, is 66 bbl-miles as advertised. As his business flourished, John was able to buy more drays, expand the storage cellars of other pubs, and thus switch to a more efficient distribution system. You may be able to work out from figure 6.9(b) that, starting from the Blue Lion and following the route shown, his distribution costs reduce to 26 bbl-miles, and reduce further if he starts from the Blue Boar. For larger networks, the costs and savings are magnified, as shown in the figure. So, careful choice of route saves money. This applies widely, and not just in the beer world. The problem of optimum routes and distribution networks has been investigated by many applied mathematicians over many years, such is its importance.

DISPENSING (CELLAR TO GLASS)

Our pint of beer has been packaged at the brewery and distributed to your favorite bar, and now you have ordered it. How is beer dispensed from the barrel (or, much more likely these days, the metal keg) in the cellar to your glass?

If a barrel is expected to be drunk in a short period, it may be placed on or above the bar, and tapped, so that the beer is poured out under gravity. This method was widely applied in England until a few decades ago and can still occasionally be seen. The beer will become warm in the

Figure 6.10. Beer drays are purely for show these days, but in yesteryears beer really was distributed from brewery to pubs in drays such as these, pulled by powerful Shire horses. These drays represent two English breweries, Thwaites and Robinson's. My thanks to David Webster for these images.

Figure 6.11. An Australian dray. Many breweries across the world pride themselves on these rigs, and much effort goes into dressing the horses as well as the draymen. Thanks to Susan Weir for this image.

bar, and this is not good news, so gravity dispensing is appropriate only for beer that is pre-chilled or is meant to be served at cellar temperature (which is several degrees below room temperature), and that is dispensed quickly. Up until the eighteenth century, beer would be poured into jugs in the cellar and carried upstairs to the patrons. This process becomes unwieldy in large taverns, with many patrons seeking to slake their thirst. In 1785, in London, Joseph Bramah invented the *beer engine,* which is a system of pipes, or *lines,* conveying beer upstairs, drawn by muscle-powered pumps.[9] In the bar, these hand pumps would be pulled, and beer raised up from the cellar in this manner by buxom barmaids.[10]

9. Bramah was something of a polymath who invented many useful devices and is perhaps best remembered for designing a burglar-proof lock. He also invented a fire engine, an hydraulic press, and an improved "water closet."
10. Made more buxom as a result of the exercise of "pulling pints," so bar-room chat informs me. Not being an expert on matters physiological, I can but repeat this much-discussed topic of conversation.

Figure 6.12. An array of Belgian beer fonts. *Inset:* Hand pumps in a Scottish pub. *Main image:* Photo from Wikipedia (John White); *inset:* image courtesy of Kirsty McFadzean.

Such hand pumps are still a fairly common sight in English pubs, but inevitably technology is taking over, with electric pumps replacing hand pumps (though often with an imitation hand pump placed on top, because hand pumps look good and satisfy traditionalists). The tall beer font common all over the beer world (see fig. 6.12) may have derived its shape from that of the old hand pump.

The physics of beer dispensing is simple and need not detain us for long. It is necessary to stabilize the beer barrel before tapping it, especially for traditional cask-conditioned beers. This is because the beer contains yeast which must be allowed to precipitate, as is also the case with my bottles of homebrew. The liquid must be kept still, and this would not be possible inside a rolling barrel. So the barrel is *stillaged* before being tapped.[11] Large barrels are usually stored on their sides, with the *shive* (the wooden fitting containing the vent hole) at the top. A *spile* (wooden peg) is driven into the shive and tightened or loosened to control the rate at which CO_2 is vented or admitted. Carbon dioxide is vented during the conditioning stage, when yeast carbonates the beer in the barrel, and is admitted when the barrel is tapped. Otherwise, a partial vacuum will build up, preventing easy pouring. In the old days, as we saw, air rather than CO_2 was admitted through the vent hole, and of course, air would spoil the beer unless the beer was consumed soon after tapping. Naturally, modern kegs, pumps, and artificial carbonation significantly simplify the whole business of dispensing beer.

Another problem that arose with the traditional methods of dispensing beer from barrels in the cellar was the uneven level of carbonation. Beer would be primed to overcarbonate in the barrel because the brewers and cellarmen knew that CO_2 would be lost as beer was dispensed. Beer would become flatter—less effervescent—as the barrel emptied simply because there was more empty space in the barrel and lower pressure, so that dissolved CO_2 would come out of the beer. Again, modern techniques of artificial carbonation and of providing CO_2 under pressure via the vent hole, solve the practical problem (but change the character of the beer—an unfortunate side effect).

DRINKING (GLASS TO STOMACH)

So there you are, sitting in a bar contemplating the freshly poured pint of beer in front of you. Centuries of human endeavor and billions of dollars of corporate research have gone into this moment (OK, and a trillion other such moments). The beer has been brewed, distributed halfway around the world, and then dispensed into your glass.[12] Now you are faced with the enormously complex task of successfully negotiating the container full of beer into your stomach without spilling it.

11. A stillage is simply a wooden or metal frame that holds the barrel firmly. Sometimes it includes wedges or springs that permit the barrel to be tilted as it empties.
12. There is skill in pouring beer into a glass, as we have seen. However, such skills are less and less in demand these days, sadly but not unusually. Today, as often as not the only skill required of a bartender, when asked to provide a beer, is the ability to take the cap off a bottle and hand the bottle to the customer.

This task—a pleasure, indeed—is not normally considered to be complex, but a little thought will convince you that I speak the truth. Indeed, the only reason that you have not as yet realized what a superbly tuned, complex, and sophisticated drinking machine you are, is that your automatic feedback systems are so smooth and self-regulating that they rarely bother to inform you of the wonders that they perform, day in and day out.

Let us consider the biomechanics of your actions as you pick up your glass, carry it to your mouth, and pour the amber nectar down your throat. You reach for the glass. Hand-eye coordination guides your metacarpal digits: your fingers are moved to within an inch of the glass, based upon information received via your eyes and transmitted to your brain. Your brain then translates the glass coordinates to your arm and hand which move your fingers into contact with the glass. Your hand contains 24 bones, 34 muscles in two muscle groups, and 48 nerves that provide it with a flexibility and finesse, a delicacy of articulation that is unmatched in the animal world. One quarter of your brain's motor cortex is devoted to providing this complex, smooth, and seemingly effortless articulation.[13] Muscles in your hand and forearm control finger movement as you close your grip on the glass. Sensory feedback from your very sensitive fingertips feel the pressure, and you automatically make fine adjustments to finger position so as to grip the glass without moving it, crushing it, or spilling any beer. Feedback from the fingertips to the brain provides the information that your brain needs to order the adjustment of finger and thumb positions. (The flow of information is suggested in fig. 6.13.) This simple action of reaching for your glass of beer thus represents the culmination of millions of years of evolution. No other animals except some of the higher primates possess an opposable thumb, so important for gripping, and no other animals at all possess the ulnar opposition capabilities of humans.[14]

13. There is an impressive YouTube video clip showing the articulation of a robot hand (see bibliography). It has taken engineers decades to reach this point of robotics development. This movie clip forcibly reminds the viewer of the dexterity of the human hand and how we take its wonderfully automatic action for granted.
14. Basically, the thumb and ring finger can touch, and the thumb and pinkie. I am

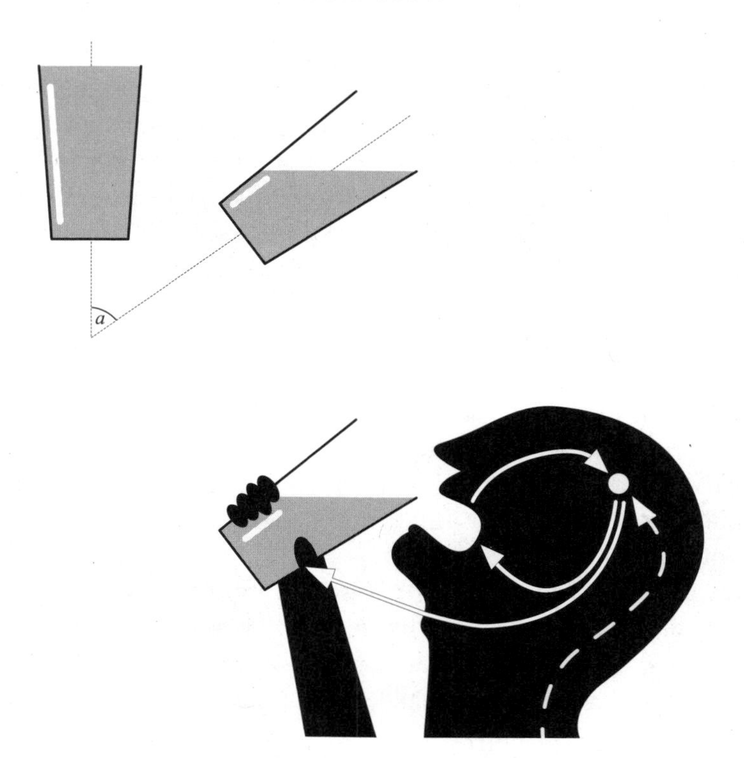

Figure 6.13. The biomechanics of beer drinking. As the glass tips, beer flows at a rate that depends in a complicated way upon the angle *a* and upon the shape of the glass. The beer-drinker and the beer glass are part of a greater whole—a complex interacting feedback system. Arrows suggest pathways for some feedback: between mouth and brain, between hand and brain, and (perhaps weakly) between stomach and brain—to tell you when you've had enough.

Now you pick up the glass. Your hand automatically weighs the glass and contents, and feedback control mechanisms adjust muscles in your hand, arm, and shoulder so that the pint is conveyed smoothly towards your mouth, while retaining the same orientation to avoid spilling a drop. The glass is brought into gentle contact with your lips; you tip it at just the right angular rate to pour beer comfortably down your ca-

---

not suggesting that this uniquely human capability arose from eons of evolution solely for the purpose of grasping pints of beer, but it is an interesting thought.

pacious gullet. More sophisticated feedback here—consider figure 6.13. The rate at which the glass angle, a, is tipped must change on the fly to maintain a constant beer flow rate. This is because of the changing geometry of glass plus beer. Of course, your brain does not direct your eyes to measure the volume of beer left and then calculate the flow rate. For one thing, the beer is too close to be seen clearly, and for another you may have your eyes closed, in an ecstasy of appreciation. Eyes open or closed, you pour the beer down without spilling any—usually. This impressive achievement results from feedback. If the mouth and throat sense that your hand, in its eagerness to pour beer, is overdoing it, then your brain is informed. Your brain tells your hand to ease off. On the other hand (as it were) if beer flow is below the optimum rate, then your hand is instructed to tip the glass faster. All this happens automatically, behind the scenes, as your mind concentrates upon the sensory pleasure of imbibing beer.

The delicate operation of conveying beer from glass to stomach can go awry because of, ironically, the effect of the beer on your nervous system. Your stomach may send only a weak signal to your brain, saying that it is full, or your brain may tell your stomach to mind its own business because your team has just scored and you are *not* going to stop drinking beer for some time yet. Whatever the reason (if reason has anything to do with beer drinking), it is not uncommon for beer drinkers to drink more beer than is strictly good for their automatic feedback systems. At that point the beer glass and lips may not dock quite as gently as they should, as smoothly as two spaceships or an ocean liner majestically arriving at port. The flow of beer may lack the precise judgment or execution that we have come to expect. Indeed, the entire motor system may display symptoms indicating a certain lack of fine coordination.[15] An inebriated man (and most of the drunken people that I have seen are male) staggers about with his clothes and limbs awry, "three sheets to the wind" as the colorful nautical phrase has it.

15. Readers may find it difficult—nay impossible—to believe this, but your author is not wholly inexperienced in such physiological states as are being described here, particularly during his student days in Scotland. Where the Eskimo has 20 different words for "snow," the Scotsman has as many words for "drunk," including *bevvied, blootered, buckled, fou, guttered, legless, moroculous, mortal, pie-eyed, pished, plastered, rat-arsed, scuppered, smashed, smeekit, steaming, stocious,* and *wrecked.*

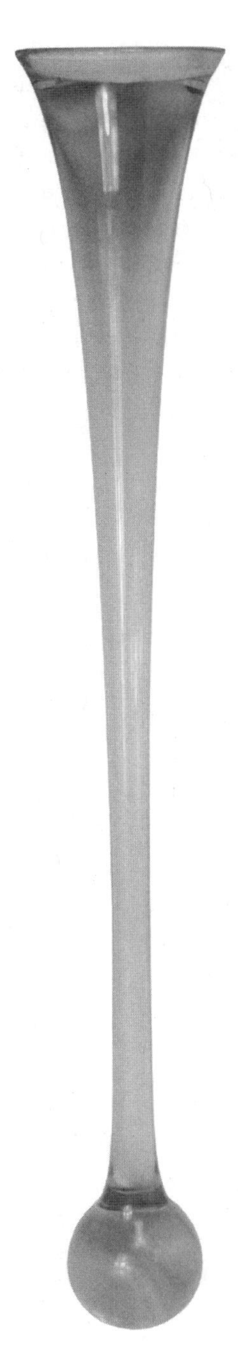

Figure 6.14. A yard of ale. Thanks to Trafford Gordon of www.drinkstuff.com for this image.

His eyes and hair are wild; he exhibits beer stains all down his shirt and trousers: in short, his feedback systems are not functioning adequately.

It is interesting to note that several drinking games intentionally test the drinker's ability to hold onto his feedback mechanisms even in the most adverse of alcoholic circumstances. Thus, for example, I once observed a student in the union bar at Edinburgh University drink a pint of beer while standing on his head. This is a difficult feat even when sober, which this gentleman was not. Quite apart from the extra feedback involved in balancing on his head using only one arm for support (the other holding his pint) he had to be very careful in bringing the beer to his lips, since any spilled beer would go up his nose. When he had successfully completed his task, an appreciative, knowledgeable, and inebriated audience applauded.

A better-known test of feedback capabilities is the drinking of a *yard of ale*. A yard of ale is a strangely shaped container (see fig. 6.14) that is indeed about a yard long and holds between two and four pints of beer. The idea is to drink the contents in one go, as quickly as possible, without pause and without spilling any beer. For the experienced beer guzzler, the difficulty in achieving this end is not the volume of beer but the "without spilling any" stipulation. When emptying the yard, a drinker has to tip up the end, and at a certain angle the beer in the bulbous end suddenly sloshes down and surges towards the drinker's face like a tidal wave. At this point, the drinker is doomed to failure. He (I have never seen a female attempt to drink a full yard of ale) may manfully open his gape and hope to swallow most of the tidal wave, but some of it will go down the wrong way, some of it will go up his nose, and a lot of it will soak him from head to foot. It seems that the trick is to spin the yard about its long axis as you drink from it—a task requiring more than the usual amount of dexterity. The resulting beer vortex causes the beer to emerge gradually from the bulb, instead of surging out. Another method you might try (should the urge to down a yard of ale come upon you), one that worked for me but won't win any prizes because it is so slow, is to rapidly lower the bulbous end once the tidal wave begins, so that beer arrives at your mouth at a rate you can deal with.

Seven

Final Thoughts

There is nothing which has yet been contrived by man, by which so much happiness is produced as by a good tavern or inn.

—Samuel Johnson (22 March 1776)

. . . and I will make it a felony to drink small beer.

—William Shakespeare (Henry VI, Part 2)

BEER QUESTIONS

The evolution of brewing from the eighteenth century is an example of the application of science to everyday life. The industrial revolution influenced brewing in no small measure: steam engines and coal led to pale ale; understanding yeast biology led to different types of brew; refrigeration revolutionized beer distribution. The product (beer) has been improved through understanding how it is made, and this is progress—and science at its best. Unfortunately, some increased understanding has also been applied in such a way as to reduce beer flavor (e.g., through filtering and pasteurization) while at the same time improving quality control; this is a sign of the times, if not of progress—but it is still applied science. The differences between large- and small-scale beer production has been a theme of my book, and these differences can be understood scientifically. You, the consumer, will already have decided, before you picked up this book, whether you prefer MB beer, or μb beer, or even nanobrewed (i.e., homebrewed) beer. You probably based your decision upon cost, taste, and availability and not upon the underlying science. Learning something of the science is fun, however, and it may lead you to some insights.

In this vein, consider the following questions (to be discussed with friends in the pub, over several beers):

- Based upon what you have learned about the history and science of beer, how do you explain the differences between the beers produced by MBs in Germany and in the United States?
- Cask-conditioned or bottle-conditioned beer would appear to be the answer to maintaining draft beer freshness and quality while distributing beer across a large network. If so, why has this approach not been more widely adopted?
- The craft brewing explosion of the last 30 years has spread widely across the United States, and yet it began in England. How is this connected with the English preference for top-fermented brews, served at cellar temperature?
- What inroads have light beers made into the heartlands of the Czech/German and the Belgian brewing traditions?

On a smaller scale, I find that thinking about beer-making scientifically helps a little bit with homebrewing. Thus, I recognize that there is a difficulty in obtaining a uniform temperature throughout the grist when it is mashing, and so I factor this into my choice of mashing temperature. I understand why yeast cells should not replicate too many times while they are fermenting a brew, and so I boost the amount of yeast pitched beyond the conventional levels. Most importantly of all, I look critically at the methods that I have been applying and seek to improve them. I let the beer drop into a carboy following primary fermentation, rather than siphon it in as the reference books tell me to do. I choose not to introduce a cold break, having understood the consequences, and the results work for me. I have experimented with the amount of sugar added as priming so as to obtain just the right degree of carbonation. Homebrewing is all about brewing beer to suit the brewer, and this can involve more than simply playing around with different recipes. I find, and you will too, that by careful scientific analysis (either theoretical or, more likely, experimental) I can improve my technique of brewing beer and also can pare down the process to minimize the considerable amount of work that homebrewing involves.[1]

BEER SAMPLING

You may think that you know how to drink beer. It is a simple process: one fills a receptacle, brings it to one's mouth (hollow side up), and drains the contents. I also entertained similar vanities until I read the detailed advice of several experts, at which point I realized that most of us are callow neophytes, by comparison. Permit me to enlighten you, and so to enhance your appreciation of beer.

I begin with the poured glass. We have seen already that there is a technique to pouring, but here I will assume that the act has been performed correctly, and with some panache, as you are demonstrating the refined art of beer sampling to your friends. You first hold up the glass, noting the cool but not frozen temperature, and cast a questioning eye on the contents: color, clarity, and head. The color should be uniform; the liquid should (except for wheat beer) be crystal clear, even without the regrettable practice of filtering. The frothy head should be full, composed of uniformly small bubbles, and it should stick to the side of the glass. Satisfied with the visual appearance of this beer, you move on to test its olfactory appeal. Six hundred and fifty aromatic compounds titillate your nose, and this is just the beginning. You now swirl the beer around in the glass to release the secondary aroma, which you breathe in with eyes closed, to improve concentration. Swirling will also test the head retention. You sniff again and breathe in through your mouth. Is the bouquet balanced? Are there any off-flavors? Only now do you move toward tasting the beer. Froth first, and then the beer, noting the difference in hop content between the two. Any sweetness in the beer is sensed at the tip of your tongue; sharpness, astringency, sparkle, and alcohol content are sensed at the sides; bitterness at

1. Homebrewing beer from grain—the so-called *full-mash* approach—is declining in popularity. This is due mainly to improvements in the quality of homebrew beer kits, which offer less flexibility and choice but are considerably more convenient. Kit beer quality is lower than full-mash brew, and the finished product costs more dollars per liter, but also costs many fewer hours. In today's busy world, time is often more important than quality or price.

the back. You exhale and sense the *mouthfeel*—the body of the beer. Finally you swallow the beer, after letting it linger in the mouth, washing over your entire palate. Once the beer is swallowed, any bitterness in the *finish*—the aftertaste—becomes apparent.

Posing? Well, perhaps, but it sure as hell beats glugging down a Lyte straight from the can.[2]

FUNNY BEER

Humor and beer go well together, and more than likely, they always have. Here is one example of many. Consider these names of homebrew organizations: *Wort Hogs, Maltose Falcons, Foam Rangers, Keystone Hops, San Andreas Malts, Los Alamos Atom Mashers, The Draught Board,* and *Quality Ale and Fermentation Fraternity* (QUAFF). Apart from a keen desire to learn about and brew good beer, the admirable folk who constitute these worthy organizations seem to share a penchant for bad puns—real groaners. Humor and math, on the other hand, may seem strange bedfellows to some people. Certainly, the technical physics papers that I have written from time to time have been long on math and short on humor—indeed, I have managed to squeeze only one joke into my technical publications, to the best of my recollection. This lack of humor in physics journals is appropriate because these journals take themselves very seriously and also because journal space is at a premium. The overworked editors cannot afford for contributing authors to waste several lines telling anecdotes. Inevitably, the result is dense math and a dry read.

In this book I have mixed two subjects that do not normally go together—math and beer are like garlic and ice cream. Should such a book be dry or humorous? At the risk of giving a false impression, I have gone for humor (possibly dry humor). It is important to understand that my analysis is no less valid because I have presented it along with a

2. I note here, incidentally, an effective though often unpleasant way to tell the difference between good beer and indifferent beer. Pour out half a glass, and let it go flat and warm up to room temperature. Now taste it. Good beer will still taste like beer, though it will not be refreshing. Bad or indifferent beer will show itself through a metallic aftertaste or intensified off-flavors. Without carbonation and the numbing effects of refrigeration, indifferent beer is left naked and revealed in all its horror.

bad pun or two, just as the history presented in chapter 1 holds true even though the chapter also contains a mythical fable. I assume that you had no difficulty understanding that Adipose Al et al. were not part of the evolution of brewing, except as a humorous illustration of the declining quality of mass-produced beer ("Macroswill"). My explanations of yeast population dynamics, homebrewing thermodynamics, bubble physics, and beer distribution networks are all based on solid physical and mathematical principles that an undergraduate physics or math student should be able to recognize. For those readers who do not have a background in these subjects, the presentation has been arranged so that the serious physics can be grasped without the math, and along with the jokes. If you have learned some physics here, as well as gained insight into the world's favorite tipple (see fig. 7.1), and maybe even laughed, then I have succeeded.

Figure 7.1. There is no doubt that people all over the world like beer. Some like it a lot. Here, a German is introduced to a glass of beer. It is obvious that these two are going to get along just fine. Photo courtesy of the Bavarian Brewers Federation, Munich, Germany.

Glossary

Acetaldehyde. A fermentation by-product that can impart a flavor or aroma of green apples to beer.

Aerobic respiration. Here, yeast respiration in the presence of oxygen.

Air lock. A device that permits gas to escape from beer in a sealed container but does not let air into the container.

Ale. Here, beer that contains little hops. A term used to describe beer produced in past centuries before the action of yeast was understood.

Alpha acids. The acids that are extracted from hops to add bitterness and aroma to beer.

Altbier. A type of dry lager from Düsseldorf.

Amber. A top-fermenting beer made with caramelized malt.

Anaerobic respiration. Yeast respiration in the absence of oxygen.

Antibubble. A hollow sphere of gas surrounded inside and out by liquid.

Attenuation. The reduction of wort specific gravity due to fermentation. A *high attenuation* of the wort means that almost all the fermentable sugars have been processed by the yeast.

Barrel (bbl). A wooden container traditionally used for transporting beer. Contains 36 imperial gallons (45 U.S. gallons).

Beading (creaming). The process of foam formation as bubbles rise out of a poured glass of beer.

Beer. (1) A mildly alcoholic beverage brewed from fermented cereals. (2) The elixir of life.

Beer engine. A mechanical pump for conveying beer up from cellar to bar.

Bitter beer. English top-fermented beer, heavily hopped and usually dark.

Black malt. Malt that has been roasted until it is black. Little fermentable starch remains; it is used for coloring and flavor. *Chocolate malt* is somewhat less roasted.

Bockbier. A strong winter lager from Munich.

Bright. A term used to describe clearness and transparency in beer.

Carbonation. The process of dissolving carbon dioxide in beer, under pressure, so that the beer effervesces when the pressure is released.

Carboy. A large glass or plastic container with a narrow neck; the opening is sealed with a bung and an air lock.

Cask-conditioned. The term used to describe beer that matures and carbonates in the barrel.

Catabolism. The multistep process by which yeast cells break down sugars.

Chill haze. A slight cloudiness that develops in some beer when it is refrigerated.

Cold break. A sudden reduction in wort temperature, intended to cause suspended proteins to precipitate out.

Conditioning. Carbonating beer naturally, as part of the maturation process.

Craft beer. Beer that is brewed for flavor.

Crystal malt. Malt that has been lightly roasted (to add body and flavor to beer).

Diacetyl. A fermentation by-product that can impart a buttery flavor or aroma to beer.

Disaccharide. A sugar, such as maltose, consisting of two glucose molecules.

Disproportionation (Ostwald ripening). The process whereby small bubbles lose gas to contiguous large bubbles, owing to pressure equalization.

Draff. Grain remaining after mashing, once wort has been extracted.

Drainage. The loss of liquid beer from the head, or froth, due to gravity.

Dray. A heavy wooden cart without sides used for haulage. Historically, beer drays were pulled by shire horses.

Dry foam. Foam that is formed from polyhedral bubbles and is stiff, with no liquid between bubbles.

Dunkel. A malty, dark lager from Bavaria.

Ester. A fermentation by-product that can impart a flavor or aroma of bananas to beer.

Fermentation. The process whereby yeast cells convert sugar into carbon dioxide, water, and alcohol.

Fermentation lock. See Air lock.

Filtration. Mechanical filtering of beer to remove all suspended solids, including yeast cells.

Finings. Material such as isinglass, Irish moss, or gelatin that is added in small quantities to beer to aid in the sedimentation of suspended solid particles.

Fobbing. Excessive foaming of beer as it is poured.

Free house. An English pub that is not owned by a brewery and that sells beer from any brewery it chooses.

Fruit beer. (1) A style of Belgian beer consisting of lambic plus fruit. (2) Any beer brewed with a significant amount of fermentable fruit.

Fusel alcohol. A fermentation by-product that can impart a flavor or smell of solvent to beer.

Grain. Cereal seeds (mostly barley) that contain fermentable starches.

Green beer. Young beer, not yet carbonated.

Grist. The mixture of grains that constitute the main source of fermentable material for a batch of craft beer. The fermentable material for beer of lower quality will include significant amounts of other ingredients, such as sugar and rice.

Gueuze. A blend of young and old lambic; perhaps the best-known beer from the Belgian tradition.

Head retention. The ability of the head on a glass of beer to remain intact and not disperse with time.

Helles. A dry, blond Munich lager.

Homebrew. Domestic craft beer.

Hops. The female flowers of the hop plant, containing acids that make beer aromatic and bitter.

Hydrometer. A homebrewer's device for measuring the density of wort.

India Pale Ale (IPA). A nineteenth-century English pale ale that was brewed strong and heavily hopped for transportation to India.

Infusion. A homebrewing method whereby grain is steeped in warm water to extract starch, forming the wort.

Kölsch. A pale ale from Cologne.

Lager. Beer produced by bottom-fermenting yeast and matured in cold storage for several months.

Lag phase. The initial phase of fermentation after the yeast has been pitched when the brew appears to be inactive.

Lambic. Flat Belgian beer that is spontaneously fermented by wild yeast. Produced with a wide range of flavors.

Macrobrewery (MB). Here, a very large brewery that brews beer for wide distribution, usually worldwide.

Macroswill. High-tech beer, produced in large quantities, for which production logistics take precedence over flavor.

Males. That subset of humanity that is fascinated by beer bubbles.

Malt. Grains that have been allowed to germinate and are then dried to terminate growth. Such treated grains form the basis of brewer's grist. Known as *pale malt* when not roasted. The process is known as *malting*.

Malthusian growth. The exponential growth of a population in ideal conditions (such as yeast after the lag phase of fermentation).

Maltose. The most important sugar that is present in wort and fermented by brewer's yeast.

Mashing. The process of steeping cracked cereal grains in warm water to extract fermentable starches.

Microbrewery (μb). Here, a small-scale brewery (one producing fewer than 25,000 bbl per year) that makes craft beer.

Mild. An English dark ale with cereal adjuncts.

Mouthfeel. The perceived body of a beer that is being tasted.

Nanobrewery (nb). Here, a domestic brewing facility that produces craft beer for the brewer's consumption, not for sale.

Nucleation site. An irregularity upon which gas dissolved in a liquid can form bubbles.

Off-flavor. Unintended beer flavor that results from imperfect fermentation owing to the presence of certain protein breakdown products.

Original gravity (OG). A brewer's measure of wort density, defined as one thousand times the specific gravity.

Pale ale. A pale, top-fermented beer that is strongly hopped.

Pasteurization. Rapid heating of beer to 70°C for 15–30 seconds to kill germs.

Pils (pilsner). Originally, a pale hopped lager from Plzen in Bohemia. Today this name is misapplied to many imitation lagers produced all over the world.

Pitch. (1) Adding brewer's yeast to wort. (2) The wax or resin lining traditionally applied to a beer barrel.

Polysaccharide. Any of various sugar molecules consisting of more than one glucose molecule bound together by water molecules.

Porter. Originally a strong and dark top-fermented beer from London, obtained by mixing young and old brown ale.

Pub. A public house or tavern that dispenses beer. Also known as heaven.

Reinheitsgebot. A set of Bavarian laws designed to ensure that beer is brewed to a high quality by limiting the permitted contents. Extended to other parts of Germany, these beer purity laws have had widespread consequences.

Shive. A wooden (nowadays plastic) fitting at the top of a beer barrel in service, containing a hole to vent carbon dioxide and admit air.

Skunking. The development of off-flavors and odors in bottled beer due to interaction with light.

Sparging (lautering). Extracting more ingredients from grist or hops by adding hot water after the grist or hops have been separated from the wort.

Spile. The wooden plug that fills the hole in a *shive,* regulating gas flow to and from a beer barrel in service.

Stillage. A scaffold which prevents movement of a beer cask in service.

Stout. Very dark, full-bodied hopped beer, a derivative of porter. Nowadays universally associated with Guinness, though there are other dry and sweet stouts.

Stuck fermentation. A fermentation that is interrupted when the yeast cells become inactive because the environment is not conducive to the process, as, for example, when the temperature is too low or the wort contains insufficient oxygen.

Surfactants. Protein molecules in beer that make it sticky and influence bubble behavior.

Three-tier system. The mandated system of beer distribution in the United States, with wholesalers separating breweries from retail outlets.

Tied house. A pub in England that is owned by a brewery and therefore sells beer only from that brewery.

Trappist. A very strong, top-fermented, bottle-conditioned beer brewed by monks. Part of the Belgian tradition of brewing.

Trub. The undesirable sediment that results from boiling the wort.

Weissbier (Weizen). Cloudy Bavarian wheat beer.

Wet foam. Fluid foam that is composed of spherical bubbles suspended in a liquid.

Wheat beer. Beer brewed from grist that contains a significant portion of wheat.

Widget. A device for creating foam with small bubbles when beer is poured from a can.

Witbier. Unfiltered Belgian wheat beer, often with orange peel or coriander. Also known as *biere blanche.*

Wort. The starchy liquid that results from mashing grist in warm water.

Yard. A long, oddly-shaped glass used in beer-drinking games (for several centuries—the yard of ale is not a recent invention).

Yeast. Microscopic single-cell fungi that are responsible for fermentation.

Zymology. The study of fermentation.

Bibliography

Alexander, Andrew, and Richard Zare. *Do Bubbles in Guinness Go Down?* http://www.chem.ed.ac.uk/guinness.

This Web site on Guinness bubbles discusses how bubbles can fall as well as rise.

Bamforth, C. W. "The Relative Significance of Physics and Chemistry for Beer Foam Excellence: Theory and Practice." *Journal of the Institute of Brewing and Distilling* 110 (2004): 259–66.

Baxter, E. D., and P. S. Hughes. *Beer: Quality, Safety, and Nutritional Aspects.* London: Royal Society of Chemistry, 2001.

BBC News. "Maths Cracks Beer Froth Mystery." 26 April 2007. http://news.bbc.co.uk/2/hi/science/nature/6592693.stm.

"Beer." *Encyclopaedia Britannica.* CD 98 Standard Edition. 1998.

BeerHistory.com. *Beer History Library.* www.beerhistory.com/library.

Brewers' Association. *Beertown.* www.beertown.org.

The Brewers' Association Web site, with definitions of "craft brewing" and much information about craft brewing in the United States.

Briggs, D. E. *Brewing: Science and Practice.* Cleveland: CRC Press, 2004.

Burch, B. *Quality Brewing.* San Rafael, CA: Joby Books, 1974.

Coghlan, Andy. "The Flight of a Beer Bubble from First Principles." *New Scientist,* 9 Nov. 1991, 24.

Dorbolo, Stéphane. "Fluid Instabilities in the Birth and Death of Antibubbles." *New Journal of Physics* 5 (2003): 161.1–161.9.

Intriguing images of bursting antibubbles, which look uncannily like jellyfish.

Dornbusch, Horst. "Beer: The Midwife of Civilization." *BeerAdvocate,* 27 Aug. 2006. http://beeradvocate.com/articles/673.

Durian, Douglas J., and Gregory A. Zimmerli. *Foam Optics and Mechanics.*

NASA/TM-2002-211195. Cleveland: NASA John H. Glenn Research Center, 2002.

Enkerli, Alexandre. "Brewing Cultures: Craft Beer and Cultural Identity in North America." Paper presented at the joint conference of the Association for the Study of Food and Society (ASFS) and the Agriculture, Food, and Human Values Society (AFHVS), Boston University, June 8, 2006. Available online at http://enkerli.files.wordpress.com/2006/06/Brewing Cultures.pdf.

Finberg, H. P. R. *The Formation of England, 550–1042*, chap. 5. London: Paladin, 1976.

German Beer Institute. *German Beer Institute: The German Beer Portal for North America*. www.germanbeerinstitute.com/index.html.

An excellent and extensive resource on all aspects of German beer (in English).

Goldammer, Ted. *The Brewers Handbook*. Clifton, VA: Apex, 2000.

Hackbarth, James J. "Multivariate Analyses of Beer Foam Stand." *Journal of the Institute of Brewing and Distilling* 112 (2006): 17–24.

Hepworth, N. J., et al. "Novel Application of Computer Vision to Determine Bubble Size Distributions in Beer." *Journal of Food Engineering* 61 (2004): 119–24.

Howstuffworks. "How Does the Widget in a Beer Can Work?" http://home.howstuffworks.com/question446.htm.

Improbable Research. *Improbable Research: Research That Makes People Laugh and Then Think*. http://improbable.com.

The Ig Nobel Prize Web site. Contains details of past winners, their achievements, the riotous prize-giving ceremonies, and other Looney-Tunes research.

Jackson, Michael. *World View*. www.beerhunter.com/worldview.html.

The extensive and authoritative Web site of this recently deceased beer guru on all aspects of beer, with many tasting notes on some very out-of-the-way brews.

James, P., and N. Thorpe. *Ancient Inventions*, chap. 7. New York: Ballantine, 1994.

Keusch, Peter. "The Tragedy of a Nice Beer Foam Head." Lecture demonstration, Institut für Organische Chemie, Universität Regensburg, Regensburg, Germany. www.uni-regensburg.de/Fakultaeten/nat_Fak_IV/Organische_Chemie/Didaktik/Keusch/D-beer_foam-e.htm.

Fat, and the detrimental effects upon beer foam, are very clearly displayed in this German academic Web site, which shows the results of a comparative experiment.

Leike, A. "Demonstration of the Exponential Decay Law Using Beer Froth." *European Journal of Physics* 23 (2002): 21–26.

Liger-Belair, G. *Uncorked: The Science of Champagne*. Princeton: Princeton University Press, 2004.

Champagne bubbles, explained in user-friendly terms by a professional bubbleologist (to coin a word).

———, et al. "On the Velocity of Expanding Spherical Gas Bubbles Rising in Line in Supersaturated Hydroalcoholic Solutions: Application to Bubble Trains in Carbonated Beverages." *Langmuir* 16 (2000): 1889–95.

Line, D. *Brewing Beers Like Those You Buy*. Hemel Hempstead, UK: Nexus, 1978.

Lusk, Lance T. "Independent Role of Beer Proteins, Melanoidins, and Polysaccharides in Foam Formation." *Journal of the American Society of Brewing Chemists* 53 (1995): 93–103.

A dull technical paper, but informative. Articles such as this one show just how much the big brewers are interested in generating and retaining the right kind of froth for their beers.

MacPherson, R. D., and Srolovitz, D. J. "The von Neumann Relation Generalized to Coarsening of Three-Dimensional Microstructures." *Nature* 446 (2007): 1053–55.

Highly technical account of how certain microstructures (such as beer bubbles) interact and grow.

Microsoft Corp. "Beer." *Microsoft Encarta Encyclopedia,* 1993–2004.

Miller, J. "Comparison of Exotherm and Carbon Dioxide Measurements in Brewing Fermentations." *Technical Quarterly of the Master Brewers Association of the Americas* 31 (1994): 95–100.

Mirsky, S. "Ale's Well with the World." *Scientific American,* April 2007.

National Archives of the United Kingdom. "Drink: The History of Alcohol, 1690–1920." www.nationalarchives.gov.uk/events/calendar/drink.htm.

The British government maintains a Web site covering everything that is British and historical, including a section on the history of alcohol in Britain.

Picard, A. "Booze Flash! Tests Confirm That Beer Bubbles Do Fall." *Toronto Globe and Mail,* 17 March 2004.

Powell, Christopher, et al. "Replicative Aging and Senescence in *Saccharomyces cerevisiae* and the Impact on Brewing Fermentations." *Microbiology* 146 (2000): 1023–34.

Protz, Roger, and Tom Cannavan. *Beer-Pages*. www.beer-pages.com/stories/complete-guide-beer.htm.

Another Web-based resource for those who are thirsty for more knowledge about all things beer. This one provides a British perspective, from guru Roger Protz.

Realbeer.com. *Realbeer.com Web Site*. www.realbeer.com/index.php.

An educational Web site covering all aspects of beer drinking, appreciation, and brewing.

Reid, David. "Scientists Create 'Antibubbles' in Belgian Beer." *Innovations Report,* 22 Dec. 2003. www.innovations-report.com/html/reports/physics_astronomy/report-24311.html.

This Web page explains antibubbles in nontechnical terms and provides a recipe for making them.

Richfield, J. "Beer Orders." *New Scientist* 2499 (May 2005): 89.

Roces, Ricardo. *Beer.* www.fordham.edu/halsall/medny/roces.html.

This extensive Web site is strong on beer history.

Saurbrei, S., et al. "The Apollonian Decay of Beer Foam Bubble Size Distribution and the Lattices of Young Diagrams and their Correlated Mixing Coefficients." *Discrete Dynamics in Nature and Society* (2006), 1–35.

The fractal nature of bubbles is presented.

Shafer N. E., and R. N. Zare. "Through a Beer Glass Darkly." *Physics Today* 44 (1991): 48–52.

Shales, K. *Brewing Better Beers*. Andover, UK: Amateur Winemaker Publications, 1967.

Shearer, D., and R. Hill. *The Home Brewer's Companion*. London: Ravette, 1982.

Stedman, E. C., and Hutchinson, E. M. *A Library of American Literature from the Earliest Settlement to the Present Time*. New York: Charles L. Webster & Co., 1889. Available online at http://books.google.com.

Warner, J. *Craze*. New York: Four Walls Eight Windows, 2002.

Wheeler, G. *Home Brewing*. St. Albans, UK: CAMRA, 1993.

Whitaker, T. "Beer Paper Wins Ig Nobel Physics Prize." *Physics World* 15 (2002): 9.

Wikipedia. *Beer*. http://en.wikipedia.org/wiki/Beer.

An extensive Web site on the character and production of beers from all nations.

Wilson, R. G., and T. R. Gourvish, eds. *The Dynamics of the International Brewing Industry Since 1800*. London: Routledge, 1998.

Witheridge, J., ed. *The Benefits of Moderate Beer Consumption*. Brussels: The Brewers of Europe, 2004.

YouTube. "Amazing Robotic Hand." *www.youtube.com/watch?v=jMHvziAikok*.

Index